T0290876

Spectrophotometric Determination of Palladium & Platinum

This versatile resource consolidates available methods for the spectropho-tometric determination of palladium and platinum and serves as a practi-cal ready-to-use guide for those researching palladium and platinum and for those working in the field of medicinal metal complexes of the two metals. The beauty of spectrophotometric methods lies in their simplicity, convenience and easy operability, not to mention their cost-effectiveness. They can be automated easily and are thus one of the most affordable methods available.

Key Features:

- Addresses analysts from all areas of industry, research labs and postgraduate students of analytical or medicinal chemistry as well as materials science.
- Details all recently developed methods for palladium and platinum determination using spectrophotometry in a single source.
- Organized so that anyone interested in a particular method using a specific reagent can go directly to those details.
- Facilitates the development of better methods for specific conditions of a sample.

Spectrophotometric Determination of Palladium & Platinum
Methods & Reagents

Dr. Ajay Kumar Goswami
Retd. Professor,
Department of Chemistry
Mohanlal Sukhadia University
Udaipur (Rajasthan), India

CRC Press
Taylor & Francis Group
Boca Raton London New York

CRC Press is an imprint of the
Taylor & Francis Group, an **informa** business

First edition published 2023
by CRC Press
6000 Broken Sound Parkway NW, Suite 300, Boca Raton, FL 33487-2742

and by CRC Press
4 Park Square, Milton Park, Abingdon, Oxon, OX14 4RN

CRC Press is an imprint of Taylor & Francis Group, LLC

Library of Congress Cataloging-in-Publication Data
Names: Goswami, Ajay Kumar, author.
Title: Spectrophotometric determination of palladium & platinum : methods & reagents / Dr. A.K. Goswami, Retd. Professor, Department of Chemistry, Mohan Lal Sukhadia University, Udaipur (Rajasthan), India.
Description: First edition. | Boca Raton : Taylor and Francis, 2023. | Includes bibliographical references and index. |
Identifiers: LCCN 2022038110 (print) | LCCN 2022038111 (ebook) | ISBN 9781032232447 (hardback) | ISBN 9781032232461 (paperback) | ISBN 9781003276418 (ebook)
Subjects: LCSH: Palladium—Analysis. | Platinum—Analysis. | Spectrophotometer.
Classification: LCC QD181.P4 G67 2023 (print) | LCC QD181.P4 (ebook) | DDC 546/.636—dc23/eng20230106
LC record available at https://lccn.loc.gov/2022038110
LC ebook record available at https://lccn.loc.gov/2022038111

ISBN: 9781032232447 (hbk)
ISBN: 9781032232461 (pbk)
ISBN: 9781003276418 (ebk)

DOI: 10.1201/9781003276418

Typeset in Palatino
by codeMantra

Dedicated

---

to
my
Late Ma and Pitaji.

Contents

Section B: Spectrophotometric Determination of
Platinum – Reagents and Methods

Chapter 7 Spectrophotometric Determination Methods
for Platinum

Preface

The present book is my essence of experience that I had during my 35 years of work in the field. I have always found dearth of single-source information for determination of individual transition metals. The need to write a single book on similar metals thus was the origin of such a monograph. Earlier, I had written on Fe and Cu and, now this book is on two important metals – Pd and Pt. I have tried to collect the recent advancements published during the past almost 10–15 years, yet I cannot claim that these are the most recent. Since the volume of publication on the methods and reagents is large, I had to limit it to the most recent ones. I am hopeful that this monograph would serve as a helpful tool to those working in the area of quantitative estimation of Pd and Pt. I have tried to be as precise and simple in writing, yet only readers would judge how far my attempt has been fruitful. Any suggestions would help me improve when writing future books.

Acknowledgements

I would fail in my duty if I do not acknowledge the few persons who have been closely associated with me during the entire process of creating this book. My wife Mrs. Diwa Goswami deserves sizeable thanks as she always created situations for stimulating my writing skills. The space and time I could snatch from her was the formation of my book. My sons Spandan and Bayar – two pillars of my strength – equally deserve a word of appreciation and acknowledgement. My student Dr. Prabhat Baroliya deserves equal thanks and acknowledgement. Last but not the least are my parents to whom I dedicate this book and who were the ones who moulded in me the skill of storytelling and expressing things in a logical manner.

Author

Prof. Ajay Kumar Goswami is an erudite scholar in the field of analytical chemistry and particularly analytical reagents and spectrophotometry. He has taught coordination chemistry and application of metal complexes to medicinal chemistry and has an experience of more than 35 years both in India and abroad. He has published more than 150 research papers in journals of impact and has a list of 30 Ph.D. students to his credit. He has been a UGC BSR Faculty Fellow and has contributed to various academic and administrative positions of the university. He has published more than ten books on various topics of graduate and postgraduate curricula and has recently published three books with renowned publishers such as CRC, De Gruyter. He is an extremely knowledgeable professor, writer and researcher.

Introduction

In chemistry, an analytical reagent is a chemical compound, the addition of which causes a chemical reaction and is essential for testing. Use of spectrophotometric reagents to quantitatively determine inorganic compounds and transition metals is one of the oldest methods, still in use despite more sophisticated methods developed and available. Precision and accuracy are two buzz words in analytical chemistry, yet cost-effective spectrophotometry has not found replacement as it incorporates both within a discipline. Expansion, industrialization and scientific development in instrumentation have brought about numerous sophisticated as well as precise analytic methods, yet the spectrophotometric method using analytical reagents has its own place in analytical chemistry. This book has brought forward spectrophotometric determination methods for platinum and palladium.

DOI: 10.1201/9781003276418-1

chapter one

Spectrophotometric Determination of Palladium and Platinum
Methods & Reagents

1.1   Introduction

Named after the asteroid Pallas, palladium (Pd) is included in the platinum group metals. The group comprises platinum, rhodium, ruthenium, iridium and osmium. Palladium is a silvery white metal with atomic number 45 and atomic weight 106.42. Along with its congener platinum, palladium contributes to making catalytic converters to convert ≃90% of the harmful hydrocarbons, carbon monoxide and nitrogen dioxide, from automobile exhausts. It is also a key component of fuel cells.

Platinum (Pt), with atomic number 78 and atomic weight 195.08, is an equally important metal from the group. A dense, malleable and ductile noble metal with silverish white colour, platinum derives its name from the Spanish term "platino" meaning "little silver". Platinum is a rarer element with six naturally occurring isotopes and an average abundance of 5 µg/kg in the Earth's crust. South Africa produces about 80% of the world's platinum. With a remarkable resistance to corrosion, it is the least reactive metal even at high temperatures.

Spectrophotometry is one of the simplest methods of analysis based on measuring the intensity of light passing through any chemical substance which absorbs light. The measurement is used to determine the amount of a known substance. It is one of the most economical and affordable tools for quantitative measurements based on transmission or reflection properties of any material as a function of wavelength.

An analytical reagent in chemistry is a chemical compound that causes a chemical reaction whose addition is essential during testing. Use of spectrophotometric reagents to quantitatively determine inorganic compounds and transition metals is one of the oldest methods, still in use in spite of more sophisticated methods developed and available. Precision and accuracy are two buzz words in analytical chemistry, yet

DOI: 10.1201/9781003276418-2

cost-effective spectrophotometry has not found replacement as it incorporates both within a discipline. Expansion, industrialization and scientific development in instrumentation have brought about numerous methods of sophisticated as well as precise analysis, yet the spectrophotometric method using analytical reagents has its own place in analytical chemistry.

The argument is well supported and evidenced by research papers appearing on these methods which have good impact. The present book is an attempt to establish this fact. The earlier book [1] on the two transition metals copper and iron was our first attempt, and now because of the dearth of a consolidated monograph on palladium and platinum, two most important industrial metals, this book is our humble attempt. Based on an exhaustive survey of literature, the book comprises different available published methods for determination of palladium and platinum during the last 20 years or so. The book hopefully would serve as a ready-to-use manual or guideline for those working in the area of palladium and platinum both in academia and in industry. The book would be equally useful for those in the pharma industry working in the area of application of metal complexes of these two metals. It would be a judicious balanced guideline for particular applications be it in analysis or synthesis. Since the synthesis of novel metal complexes of palladium and platinum in recent years has become an extremely challenging area of medicinal chemistry, inorganic medicines too have become futuristic avenues of human health care [2].

1.2   Spectrophotometric Methods

Development of analytical chemistry has been one of the most important milestones of the modern era in chemistry. Spectrophotometry is another extensively used tool or technique developed by T.S. West. The technique has been subsequently used by analytical chemists to determine all types of samples including metal complexes. This is one of the most widely used technique for the quantitative determination of various inorganic, organic and biochemical samples. The technique is used for quantitative analysis by almost all clinical and chemical laboratories worldwide. The method is simple, selective and sensitive for determination of transition metals or coloured substances. The technique uses UV (100–400 nm) and visible radiations (400–700 nm) and is also called UV–vis spectrophotometry. When these radiations interact with the sample under investigation, it causes changes in the electronic energy levels within the molecules. The interaction of light by the sample is due to interaction with the electronic and vibrational modes of molecules. Every individual molecule has an individual set of energy levels associated with the type of chemical bonds and nuclei and so absorbs light of a specific wavelength. This results in unique spectral properties. Taking advantage of this, analytical chemistry applies this technique for the quantitative determination of such samples.

There are diverse methods included under the technique, such as extractive spectrophotometry, direct spectrophotometry, solid-phase extraction method, microemulsion, microscale determination, kinetic spectrophotometric determination, simultaneous spectrophotometric determination, stopped flow spectrophotometric determination, flow injection methods, phase separation and derivative spectrophotometric methods, to name a few. The method basically can be designed as per the nature of the sample and the particular requirement.

UV–vis spectrophotometry is useful in determining the presence of different compounds like transition metal ions, highly conjugated organic molecules and transition metals. The technique is based on exciting a metal's d electron from its original ground-state configuration to an excited state. The source of excitation is light. When energy in the form of photons is directed at a transition metal complex, it results in exciting a d electron by use of this energy. A UV–vis spectrophotometer thus measures the energy of excited electrons at a specific wavelength of light from the UV–vis region of light.

1.3 Theory behind the Technique

Although the detailed theoretical part can be consulted from any textbook on metal complexes and their spectrophotometric studies, a brief conceptual description is given here to apprise any reader of the basics.

a. Splitting of d orbitals

We know that d orbitals contain five types of sub-orbitals, viz. d_{xy}, d_{yz}, d_{xz}, $d_{x^2-y^2}$ and d_{z^2} as shown in Figure 1.1.

In the absence of any magnetic field, for example, when there are no electrons in all the sub-orbitals, they combine to form a degenerate spherical orbital, i.e., the orbitals of the same subshell have identical energies. Soon as electrons are introduced into this spherical singular degenerate orbital, it differentiates back into sub-orbitals, for example, when transition metal is bonded to a set of ligands. This is shown pictorially in Figure 1.2.

b. Origin of colour in transition metal complexes – crux of the spectrophotometry

The most important part of spectrophotometric studies of transition metal complexes is the origin of colour which is caused by differentiation of d orbitals in the presence of electrons. Colour in a way originates from this differentiation or we call splitting of the d orbitals

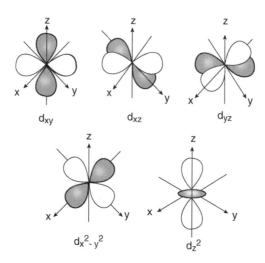

Figure 1.1 Shape of d orbitals.

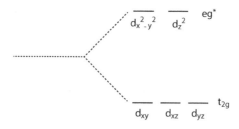

Figure 1.2 Splitting of d-orbitals.

into two sets, higher- and lower-energy orbitals. For example, an electron in the t_{2g} bonding orbital can be excited to the eg* bonding orbital by light. When this electron comes back to the ground state, some part of energy is released in the form of colour. The simple pictorial representation of the phenomenon is shown in Figure 1.3.

The colour emerges when a specific wavelength of light excites an electron to the eg* orbital on the basis of the spectrum obtained. The emission of a particular colour is directly correlated to the wavelength absorbed. Whichever colour is absorbed, its complementary colour is emitted. The phenomenon can be over-simplified by depicting it with a colour wheel. Figure 1.4 shows which complementary colour would be emitted when a particular wavelength light is falling.

When a metal complex emits green light, the opposite colour is absorbed which is red in the wheel (630–730 nm).

Figure 1.3 Electronic transition in d-orbitals.

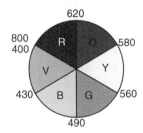

Figure 1.4 Origin of spectrum.

c. Theoretical concept behind the origin of colour in metal complexes and the ligand field theory

A complex in the simplest words is defined as "a species formed by the association of two or more simpler species each capable of independent existence" [3]. Of these two or more species, when one of the species is a metal, the resulting species is called a "metal complex". Central metal atom of a complex can be any metal, but some metal atoms can serve better as a central atom. For example, transition metals function excellently, whereas alkali metals rarely do so.

What Is a Ligand?

The term "ligand" was introduced long back with two different meanings. Sometimes it applies to the particular atom, for instance, nitrogen atom in ammonia, or an entire molecule, but essentially it means species attached to a central metal.

How Do Ligands Affect Colour?

It is understood that the atoms bonded to any transition metal influence the wavelength of any complex which is needed to emit the particular colour. This is understandably the essence of ligand field theory. In other words, the ligands play a major role in determining the wavelength of light needed to excite an electron. Thus emitted light or colour directly correlates to this. Two common terms, low spin and high spin, describe the difference in energy levels of the t_{2g} and eg* orbitals, with relation to the wavelength needed to excite an electron from a t_{2g} to eg* level in a low-spin complex; the attached ligands raise the energy of eg* orbitals so much that the ground-state configuration fills the first six electrons in the t_{2g} orbital and no eg* orbital is filled. This means that high-energy wavelength of light, e.g., violet, blue and green, is required to excite an electron to the eg* bonding orbital. This results in emission of yellow, orange and red light which are the complementary colours (opposite) in the wheel. On the contrary, a high-spin complex possesses ligands which lower the energy level of eg* orbital which means low-energy light such as red, orange and yellow or even high-energy light can excite an electron. Thus, high-spin complexes are all inclusive, while low-spin complexes are half exclusive in relation to wavelengths required to excite an electron. The simplest explanations described above are an attempt to elucidate the origin and interpretation of colour in metal complexes, specifically the transition metal complexes. Any avid reader may refer to a standard text-book on the subject.

1.4   Important Characteristics of Spectrophotometric Method

The advantages or choice of spectrophotometric methods can be summarized to highlight it as an inevitable tool to the analytical chemist.

a. Wide applicability

Spectrophotometric methods have wide applicability which is evidenced by the fact that 90% of the clinical labs use this method. The diverse use of these methods includes research labs, pharmaceutical labs, environmental labs, industry-based labs, mineral labs and food processing labs, just to mention a few.

b. Moderate to high selectivity

Selecting a suitable wavelength is important in spectrophotometric analysis to avoid making measurements in a region where the molar absorptivity changes quickly with the wavelength. Thus, it is desirable to select a wavelength where the absorbance is maximum and remains constant over a reasonable range of selection for coloured solution.

c. Good accuracy

With precautions in the analysis using UV–vis radiations, the accuracy range can be achieved to very high levels. In general, the relative error is in the range of 1%–5%, which can be improved by an analyst.

d. **Ease and convenience**

The beauty of spectrophotometric methods lies in their simplicity, convenience and easy operation not to mention their cost efficiency. These can be automated easily and are thus the most affordable methods.

1.5 Procedural Details of Spectrophotometric Methods – A Brief Outline

The first requirement of any spectrophotometric method is the development of suitable conditions which give a reproducible relationship between absorbance and concentration of the analyte under screening. The quantitative application of the method is based on the fact that the number of photons absorbed is directly proportional to the concentration of the substance. So, the methodology involves the following steps:

a. **Choice of wavelength**

In spectrophotometric analysis, the wavelength of absorbance needs to be selected in order to record measurement in a region where molar absorptivity does not change very quickly with change in wavelength. Thus, a suitable wavelength is chosen as maximum absorbance (λ_{max}) or working wavelength (which is the region of constant maximum absorbance).

b. **Colour development**

A solution or object appears coloured as it absorbs a range of radiation in the visible region of spectrum. The molecular structure is closely related to the colour of any solution of analyte. The concentration is measured by determining the extent of absorption of lights at an appropriate wavelength of absorption of the coloured solution. In spectrophotometric determination of transition metal complexes for example, formation of the complex is not instantaneous; thus, it takes a few seconds or even minutes before the final colour of maximum intensity is developed.

c. **Beer's law**

The basis of quantitative analysis using spectrophotometry is observance of Beer's law. It is the fundamental law to be followed for the determination of analyte in UV or visible region. It is described by a plot of absorbance against concentration which is a straight line passing through the origin and covers a wide range of concentrations in dissolved solution of analyte. A calibration graph is used for the determination of concentration of analyte in question.

d. Sensitivity of spectrophotometric method

The sensitivity of spectrophotometric method is described in terms of molar absorptivity (ξ) of the metal–ligand complex. The sensitivity of the method is extremely important in trace determination and thus is a characteristic taken in considering a particular method.

e. Effect of diverse ions and interference of other analytes

The effect of other ions on the determination of a particular species increases or decreases the applicability of any method. The tolerance limits of the interfering ions are established at those concentrations where not more than ±2.0% change in absorbance is observed.

f. Precision and accuracy

Precision is the reproducibility of any set of retrieved data, whereas accuracy denotes the correctness of any observed data. The terms are frequently used to define the validity of a particular method. Both of these depend on major factors like instrumental error, chemical variables or even skill of the operator. In general, spectrophotometric methods give results with a precision of 0.5%–2% under suitable measuring conditions. Any standard textbook can be consulted to understand the principle and applications of spectrophotometric methods or quantitative analysis. This book focuses on spectrophotometric methods for the determination of palladium and platinum, and thus addresses the readers or the researchers in this field using these methods.

1.6   Spectrophotometric Reagents

Organic reagents have found multiple applications in quantitative analysis, more so in the spectrophotometric determination of transition metals. This is more emphasized by their high degree of sensitivity and selectivity under a chosen set of experimental conditions. The quest for proper reagents was streamlined after the development of coordination chemistry by A. Werner and his school. The big step was further taken after rationalization of analytical methods. However, the role of organic reagents in spectrophotometric determination of transition metals is now well known and established. A lot of research is being done on the development of better reagents and their applications. For this, the ideal analytical reagent must have a proper functional group which is able to react with the metal or its compound through the donor site of the ligand. These ligands normally may have hydroxyl, carbonyl, halogen, nitrogen, oxygen, sulphur or groups containing them. It is certain that organic reagents have acquired a significant position in the area of spectrophotometric analysis in the last few decades. A list of some organic reagents being used for the spectro-photometric determination of transition metals is being given to make the reader aware of their significance:

1. Oxine (8-hydroxyquinoline)
2. Dithizone (1,5-diphenylthiocarbazone)
3. Cupferron (N-nitroso-N-phenylhydroxylamine ammonium salt)
4. Dialkyldithiocarbamates
5. Salicylaldoxime
6. Cupron
7. α-nitroso-β-naphthol
8. Nitron
9. Thionalide (2-mercapto-N-naphthyl accetamide)
10. Biacetyl dioxine
11. Complexones, e.g., EDTA, EGTA or like
12. 1,3-diphenylhydroxytriazenes

The list can be exhaustive, thus the details of developing reagents by changing functional groups are a challenge for the researchers. This challenge has brought out several new reagents in the past. The chemistry of metal complexes and their interesting biological properties has given fillip to medicinal chemistry and drug development in recent years. The successful theoretical prediction and computer-aided drug designing (CADD) has further helped the researchers in developing new ligands suited to act at a particular enzyme site with desired activity. Excellent software for this prediction and in silico activity predictions are few such tools making biologically active metal complexes and such ligands a curious area of human knowledge.

The new objective of developing drugs based on metal complexes has further grown interest in their synthesis, characterization and applications enhancing the use of spectrophotometric determination of such metals including palladium and platinum. Platinum has been one of the most explored metals owing to the anti-cancer properties of cisplatin and related compounds in medicinal chemistry. On the other hand, palladium too is equally important due to its application in medical appliances and medicines. The importance of spectrophotometric determination of transition metals including palladium and platinum can be evidenced by the voluminous publications in these areas of research. The author and his research group have published a number of single-metal reviews and books on spectrophotometric determination of different transition metals during the last 20 years, which further emphasizes their significance [4–13]. In continuation of this, the author has recently published a book on spectrophotometric determination of copper and iron [1]. This book is

Table 1.1 List of Spectrophotometric Reagents for Determination of Palladium

S. No.	Name of Analytical Reagent	Absorption Maximum (λ_{max}) (nm)	pH/Medium	Molar Absorptivity (L/mol cm)	Sandell's Sensitivity Detection Limit
			Acidic/alkaline	–	–
1	4-(4-Dimethylaminobenzyl lideneimino)-5-methyl-4H-1, 2,4-triazole-3-thiol (DMABIMTT)				
2	4-(4-Fluorobenzylideneimino)-3-methyl-5-mercapto-1,2,4-triazole (FBIMMT)	390	HCl	5.404×10^3 L/mol cm	$0.0196\ \mu g/cm^2$
3	α-Nitroso-β-naphthol	425	Alkaline		
4	Pyridoxal thiosemicarbazone (PTSC)	420	2.0	1.63×10^4 L/mol cm	$0.635\ \mu g/cm^2$
5	1-(2-Benzothiazolylazo)-2-hydroxy-3-naphthoic acid (BTAHN)	669	5.0 M (@ HCl + TMAB)	2.61×10^5 L/mol cm	Detection limit 6.3 and 19.5 ng
6	para-Methylphenylthiourea (PMPT)	300	–	0.843×10^3	$0.125\ \mu g/mol/cm^2$
7	Methylene blue (MBB)	662.4	$(NH_4)_2S_2O_8$ in citric acid buffer		$2.0\ ng/cm^3$ (DL)
8	2,4-Dihydroxy-5-iodo-α-phenylacetophenone oxime (DHI-α-PAO)	420	2.0	4.23×10^2	$0.2516\ ng/cm^2$
9	Naphthol green B	721	Acidic	–	$0.9\ ng/mL$ (DL)
10	Cefixime	352	2.6	1.224×10^4	$.002\ \mu g/cm^2/.001$ absorbance unit
11	2-(5-Bromo-2-pyridylazo)-5-[N-n-propyl-N-(sulfopropyl) amino]aniline (5-Br-PSAI)	612	–	–	–

(Continued)

Table 1.1 (Continued) List of Spectrophotometric Reagents for Determination of Palladium

S. No.	Name of Analytical Reagent	Absorption Maximum (λ_{max}) (nm)	pH/Medium	Molar Absorptivity (L/mol cm)	Sandell's Sensitivity Detection Limit
12	4-[N'-4-(4-lmino-2-oxo-thiazolidin-5-ylidene)-hydrazino]benzenesulfonic acid	438	7.5×10^3	–	–
13	2-Mercaptoethanol		4.0	2.2634×10^4	7.2154×10^{-3} µg/cm^2
14	L-cystine	369	4.0	2.69×10^4	7.89×10^{-4} µg/cm^2
15	1-(5-Benzylthiazol-2-yl) azonaphthalen-2-ol (BnTAN)	600–684	–	6.60×10^3	
16	5-Hydroxyimino-4-imino-1,3-thiazolidin-2-one (HITO)		5.0		6.0×10^{-6} to 6.0×10^{-3} M (detection limit)
17	5,6-Diphenyl-2,3-dihydro-1,2,4-triazine-3-thione	385	–	6.67×10^3	–
18	2-(5-Bromo-2-oxoindolin-3-ylidene) hydrazine carbothioamide, [5-bromoisatin thiosemicarbazone] (HBITSC)	520	0–4.0	7450	14.3 ng/cm^2
19	o-Methoxyphenylthiourea (OMePT)	325	–	3.38×10^3	0.31 µg/cm^2
20	N-Decylpyridin-4-amine in 1:1 ammonia and pyridine-2-thiol	410	–	1.9×10^5	0.065 µg/cm^2
21	Pyridoxal thiosemicarbazone (PDT)	420	2.0 (conc. HCl + CH$_3$COONa)	1.63×10^4	0.635 µg/cm^2
22	2-Hydrazinopyridine	510	–	2.978×10^3	–

(Continued)

Table 1.1 (Continued) List of Spectrophotometric Reagents for Determination of Palladium

S. No.	Name of Analytical Reagent	Absorption Maximum (λ_{max}) (nm)	pH/Medium	Molar Absorptivity (L/mol cm)	Sandell's Sensitivity Detection Limit
23	(+)-Cis 3-(acetyloxy)-5-[2-(dimethylamino) ethyl]-2,3-dihydro-2-(4-methoxyphenyl)-1,5-benzothiazepin-4(5H)-one, monohydrochloride (Diltiazem)	400	–	8.5×10^2	–
24	(NaHMICdt. $2H_2O$) sodium salt of hexamethyleneimine carbodithioate	435	0.5–2.0	0.754×10^4	0.0140 μg/mol/cm^2
25	α-Furildioxime	–	–	–	1.1 μg/L (DL)
26	3-Hydroxy-2-(2'-thienyl)-4H-chromen-4-one	455	8.5–9.2	3.30×10^4	0.0032 μg/cm^2
27	3,4-Dihydroxybenzaldehyde isonicotinoylhydrazone (3,4-DHBINH)	362	3.0–7.0	–	–
28	3,4-Dihydroxybenzaldehyde-isonicotinoylhydrazone (3,4 DHBINH)	380	3.0	0.53×10^4	0.02 μg/cm^2 (SS)
29	N,N,N',N'-tetra(2-ethyl hexyl) thiodiglycolamide T(2EH) TDGA	300	–	1.29×10^5	0.0948 μg/mL (dL)
30	p-[N,N-bis (2-chloroethyl)amino] benzaldehyde thiosemicarbazone (CEABT)	395	1–2	4.05×10^4	0.0026 μg/cm^2
31	2-Hydroxy-5-methylacetophenone isonicotinoylhydrazone (HMAINH)	385	0.01–0.015 M H_2SO_4	5.320×10^3	0.02 μg/cm^2
32	Isonitroso-p-methyl acetophenone phenyl hydrazone (HIMAPH)	470	0.0–5.0	13.305	–
33	Thioglycolic acid	–	11	2.0692×10^4	6.8287×10^{-3} g/cm^2

(Continued)

Table 1.1 (Continued) List of Spectrophotometric Reagents for Determination of Palladium

S. No.	Name of Analytical Reagent	Absorption Maximum (λ_{max}) (nm)	pH/Medium	Molar Absorptivity (L/mol cm)	Sandell's Sensitivity Detection Limit
34	4-Amino-3-mercapto-6-methyl-1,2,4-triazin-(4H)5-one	–	11	2.1121×10^4	5.3287×10^{-3} g/cm^2
35	Isonitroso-p-nitroacetophenone thiosemicarbazone (HINATS)	410	0.0–4.0	910	–
36	1-(2-Naphthalene)-3-(2-thiazole)triazine (NTT)	–	–	4.07×10^4	–
37	Dibromo-o-nitrophenyl fluorone (DBON-PF)	474	2.2	1.07×10^5	–
38	p-Antipyrineazobenzoic acid (AABA)	452	3.5–5.5	1.22×10^5	–
39	4-(N,N-diethylamino) benzaldehyde thiosemicarbazone (DEABT)		3.0	3.33×10^4	0.032 μg/cm^2
40	1-(2-Pyridylazo)-2-naphthol (PAN) fibre optic linear array detection (FO-LADS)	Spectrometry	–	–	–
41	2-Hydroxy-1-naphthalene carboxaldehyde hydrazine carboxamide	410	4.2	0.4×10^3	1.50×10^{-5} μg/cm^2
42	2,5-Dimercapto-1,3,4-thiadiazole (DMTD) + Triton X-100	375	–	3.03×10^4	3.5 ng/cm^2
43	Azure I + NaH$_2$PO$_4$	647	–	–	4.3 ng/mL (DL)
44	1-(4-Antipyrine)-3-(2-thiazolyl)triazine (ATTA)	480	9.5	5.33×10^4	–
45	4,4-Bis(dimethylamino) thiobenzophenone		3.0	–	–

(Continued)

Table 1.1 (Continued) List of Spectrophotometric Reagents for Determination of Palladium

S. No.	Name of Analytical Reagent	Absorption Maximum (λ_{max}) (nm)	pH/Medium	Molar Absorptivity (L/mol cm)	Sandell's Sensitivity Detection Limit
46	N,N-diphenyl and N,N-1,diphenylurea (Allylthioureas)		1.0–7.0 1.0–5.0	$1.6 \times 10^4 (\varepsilon 330)$	–
47	4-(2-Pyridylazo)resorcinol (PAR)	520	9.0–11.0	8.0×10^5	$0.49 \ \mu g/cm^2$
48	1,3-Bis(4-amino-3-mercapto-1,2,4-triazol-5-yl) mercaptoethane (AMTME)	380	4.5–7.0	4.8×10^3	$0.0222 \ \mu g/cm^2$
49	4-Hydroxybenzaldehyde thiosemicarbazone	359	3.0	2.68×10^4	$0.00395 \ \mu g/cm^2$
50	2,4-Dihydroxybenzophenone oxime (BHBPO)	410	4.0	0.672×10^2	$0.06729 \ \mu g/mL$
51	2-(2-Quinolinylazo)-5-dimethylaminobenzoic acid (QMDMAB)	630	4.0	1.03×10^5	–
52	Benzildithiosemicarbazone	395	2.5	3.018×10^4	$0.0035 \ \mu g/cm$
53	2-Cis 3,7-di-Me 2,5-octadiene 1-oxime	445	0.03–0.14 M ACOH	2.434×10^3	$0.007 \ \mu g/cm^2$
54	Pyridoxal-4-phenyl-3-thiosemi carbazone (PPT)	460	3.0	2.20×10^4	$4.85 \times 10^{-3} \ \mu g/cm^2$
55	Pyrogallol red (PGR) + H_2O_2	540	9.7		$0.017 \ \mu g/me$
56	2,2'-Dipyridyl-2-pyridylhydrazone (DPPH)	560	–	–	$0.007 \ \mu g/L$
57	N-octyl-N-(sodium p-amino benzenesulfonate) thiourea (OPT)	299	5.0–5.6	1.38×10^5	–
58	2-Amylthio-p-nitro-acetophenone	430	7–8 M HoAC Medium	–	–
59	4-(2-Thiazolylazo)-5-diethylaminobenzene	590	4.0	1.08×10^5	–

(Continued)

Table 1.1 (Continued) List of Spectrophotometric Reagents for Determination of Palladium

S. No.	Name of Analytical Reagent	Absorption Maximum (λ_{max}) (nm)	pH/Medium	Molar Absorptivity (L/mol cm)	Sandell's Sensitivity Detection Limit
60	α-Benzylmonoxime-molten benzophenone	545	H_2SO_4 Soln		3 mv/mL (DL)
61	5-Methyl-2,3-hexanedione dioxime (H2MHDDO)	379	0.5–1.5	3.894×10^3	
62	5-(5-Nitro-2-hydroxyphenylazo)rhodanine (DP)	474	H_3PO_4 medium	4.0×10^4	–
63	3-Phenoxybenzaldoxime	435	4.0	2.434×10^3	
64	Methdilazine hydrochloride	555	Buffer	1.66×10^4	$20.3 \ \text{ng/cm}^2$
65	2-Hydroxy-3-carboxy-5-sulfophenyldiazo aminoazobenzene (HCSDAA)	530	9.3–10.6 Na_2CO_3– $NaHCO_3$ buffer + Triton X-100	1.05×10^5	–
66	p-Dimethylaminobenzylidenerhodanine + methyl isobutylketone	2.4		3.0×10^4	0.1 mL/L
67	Rubeanic acid in the presence of ethylenediamine (EDA), diethyltriamine (DETA), DMSO and ammonium thiocyanate (ATC)	406–422 (EDA) 417–423 (DETA) 415–425 (DMSO) 408–420 (ATC)	5.0	6724 (Direct) 9966 RA-NH$_4$SCN System	–
68	N-butyl-N′-(sodium p-aminobenzenesulfonate) thiourea (BPT)	294	4.6–5.6	2.46×10^5	–
69	1-Phenyl-1,2-butanedione dioxime (H$_2$PBDD)	380	6.5–9.0	3.74×10^3	–

(Continued)

Table 1.1 (Continued) List of Spectrophotometric Reagents for Determination of Palladium

S. No.	Name of Analytical Reagent	Absorption Maximum (λ_{max}) (nm)	pH/Medium	Molar Absorptivity (L/mol cm)	Sandell's Sensitivity Detection Limit
70	Pyridopyridazine dithione (PPD)	570 615	Aqueous micellar medium	4.68×10^4 5.74×10^4	—
71	2,6-Dibromo-4-carboxybenzenediazoamino azobenzene (DBCDAA) + 1,10-phenanthroline (Phen)		$9.7–10.6$ Na_2CO_3 $NaHCO_3$ buffer	7.16×10^4	—
72	p-Nitroisonitrosoacetophenone	425	5.4–5.6	1.355×10^4	—
73	Triphenylphosphine sulphide (extractant), 1-(2-pyridylazo)-2-naphthol (reagent)	—	—	—	—
74	N,N'-Dipyridylthiourea (DPyT)	330		2.36×10^4	0.0065 µg/mL
75	1-Phenyl-3-(3-ethyl-5-mercapto-1,2,4-triazol-4-yl) thiourea	335	0.8–0.16 M NaOH	2.42×10^4	4.397×10^{-3} µg/cm²
76	Isonitrosopropionylthiophene	434	Dilute acetic acid		4.3×10^{-3} µg/cm²
77	2,6,7-Trihydroxy-9-(3,5-dibromo-4-hydroxy) phenylfluorone (DBHPF) + Cetyl trimethylammonium bromide (CTMAB)	600	6.1–6.7	9.48×10^4	1.12×10^{-3} µg/cm²
78	Nicotinaldehyde 4-phenyl-3-thiosemicarbazone (NPS)	365	3.0	2.81×10^4	—
79	2-Mercapto-4-methyl-5-phenyl azopyrimidine	—	Nitric acid	2.05×10^4	0.0052 µg/cm²
80	3-Hydroxy-3-phenyl-1-p-carboxyphenyltriazene	430	3.0–3.5	14000	7.6 ng/cm²

(*Continued*)

Table 1.1 (Continued) List of Spectrophotometric Reagents for Determination of Palladium

S. No.	Name of Analytical Reagent	Absorption Maximum (λ_{max}) (nm)	pH/Medium	Molar Absorptivity (L/mol cm)	Sandell's Sensitivity Detection Limit
81	8-Aminoquinoline-5-azo-p-benzoic acid + cetyltrimethylammonium bromide	610		7.2×10^4	
82	3-Thiophenealdehyde-4-phenyl-3-thiosemicarbazone (TAPS)		$1\ M\ H_2SO_4$		
83	α-Benzoin oxime (ABO) + Iso-Bu-MeMe-Ketone (Hexone)			4.0×10^3	
84	3-Hydroxy-3-p-chlorophenyltriazene				
85	2-Mercaptonicotinic acid (MENA)	410	$0.1\ M\ NaClO_4$ (0.5–4.5)	1×10^4	
86	N-hydroxy-N,N'-diphenylbenzamidine (HDPBA)	400	5.0 ± 0.2	6.4×10^3	0.0967 µg/cm^2
87	R + presence of 1-(2-pyridylazo)-2-naphthol	620		1.58×10^4	0.1 µg/cm^2
	5-Bromosalicylaldehyde 4-phenyl-3-thiosemicarbazone (BSPS)	412	$0.05\ M\ H_2SO_4$	1.42×10^4	
88	5,6-Dimethyl-1,3-indanedione 2-oxime	370	5.5	2.98×10^4	–
89	Pyridylazoresorcinol	535	–	–	0.0034 mg/cm^2
90	2-N-(2-mercaptophenyl)-1,2,3-benzenothiadiazoline (MBDT)	788	–	2.9×10^4	3.6×10^{-3} mg/cm^2
91	Thiomichler's ketone (TMK), 4,4-bis(dimethyl) amino)thiobenzophenone		3.0	–	–

(Continued)

Table 1.1 (Continued) List of Spectrophotometric Reagents for Determination of Palladium

S. No.	Name of Analytical Reagent	Absorption Maximum (λ_{max}) (nm)	pH/Medium	Molar Absorptivity (L/mol cm)	Sandell's Sensitivity Detection Limit
92	Anion exchange separation	407	HCl, HClO$_4$ eluents		1.3 µg/L (DL)
93	p-[4-(3,5-Dimethyl isoxazolyl)azophenylazo] calix (4) arene [DMIAPAC]	–	–	1.73×10^4	0.0061
94	3-Hydroxy-2-methyl-1-phenyl-4-pyridone	345	Aqueous sulphuric acid medium (1.5–3.0)	1.89×10^4	
95	3-Phenoxybenzaldoxime	435	4.0	2.434×10^3	–
96	p-[N,N-bis (2-chloroethyl)amino] benzaldehyde thiosemicarbazone	395	1–2	4.05×10^4	0.0025 µg/cm^2
97	α-Benzilmonoxime + Triton X-100	441.8 77.0	0.10 M HClO$_4$		0.07 (DL) 0.10 µg/mL
98	5-Hydroxyimino-4-1,3-thiazolidin-2-one (HITO)	350	5.0	5.9×10^3	
99(a)	2-(2-Quinolinylazo)-5-diethylaminobenzoic acid	–			–
(b)		628		1.43×10^5	0.02 µg/L (DL)
100	1-(2-Pyridylazo)-2-naphtholate	659 681			0.3 ng/mL

(Continued)

Table 1.1 (Continued) List of Spectrophotometric Reagents for Determination of Palladium

S. No.	Name of Analytical Reagent	Absorption Maximum (λ_{max}) (nm)	pH/Medium	Molar Absorptivity (L/mol cm)	Sandell's Sensitivity Detection Limit
101	Optical test-strip using 5(p-dimethylaminobenzylidene) rhodanine in THF				0.1 mg/L
102	Michler's ketone (TMK) 4,4'-bis(dimethylamino)thiobenzophenone	–	3.0		0.4 ng/mL
103	2-(2-Quinolinylazo)-5-dimethylaminoaniline (QADMAA)		0.05–0.5 mol/L HCl	1.41×10^5	0.02 µg/L (DL)
104	Piperonal thiosemicarbazone	363	–	3.80×10^4	2.8×10^{-3} µg/cm²
105	2-(2-Quinolinylazo)-5-diethylaminoaniline (QADEAA)	615	2–2.0		
106	5,6-Diphenyl-2,3-dihydro-1,2,4-triazine-3-thione (DPhDTT)	385	–	6.67×10^3	–
107	2-[(E)-N-(2-[(E)-[(2-hydroxyphenyl) methylidene] amino]phenyl](methyl)amino} phenyl]carboximidoyl]phenol (HHMCP)	560	9.0	0.47×10^2	–
108	2-Methyl-5-(4-carboxyphenylazo)-8-hydroxyquinoline	–	3.2	2.15×10^5	0.1 µg/L
109	Pthaldehydic acid thiosemicarbazone (PAATSC)	–	4.0 HAC/NaAC	5.1×10^4	23 ng/mL (DL)
110	N-phenylbenzimidoyl thiourea (PBITU)	345	0.2–2.0 M HCl	1.20×10^4	80 ng/mL (DL)

(Continued)

Table 1.1 (Continued) List of Spectrophotometric Reagents for Determination of Palladium

S. No.	Name of Analytical Reagent	Absorption Maximum (λ_{max}) (nm)	pH/Medium	Molar Absorptivity (L/mol cm)	Sandell's Sensitivity Detection Limit
111	2-Aminoacetophenone isonicotinoylhydrazone (2-AAINH)	500	3–5	3.00×10^4	$0.0035 \, \mu g/cm^2$
112	Crystal violet with hypophosphite	590			
113	O-Methylphenylthiourea (OMPT)	340	HCl	2.85×10^3	$0.037 \, \mu g/cm^2$
114	2-Hydroxy-3-methoxybenzaldehyde thiosemicarbazone (HMBATSC)	380	1.0–7.0	2.198×10^4	$0.049 \, \mu g/cm^2$
115	Diacetylmonoxide-(p-anisyl)-thiosemicarbazone	440	Acetic acid	3.8×10^4	$2.8 \, ng/cm^2$
116	Hexyl benzimidazolyl sulphide (HBMS) + Emulsifier–OP	452	0.01–0.1 M/L HCl	2.08×10^5	$0.1 \, \mu g/L$ (DL)
117	p-Sulfobenzylidene-rhodanine	535	2.0	7.79×10^4	–
118	3,5-Dimethoxy-4-hydroxybenzaldehyde isonicotinoylhydrazone (DMI + BIH) + Triton X-100	382	5.5	2.44×10^4	$0.0044 \, \mu g/cm^2$
119	1,3-Bis(hydroxymethyl) benzimidazole-2-thione (BHMBT) + Me isobutylketone	370		1.543×10^4	$0.0068 \, \mu g/cm^2$
120	2,6-Diacetylpyridine bis-4-phenyl-3-thiosemicarbazone (2,6-DAPBSPTSC)	410	4.0	1.156×10^4	$0.0092 \, \mu g/cm^2$
121	2-Hydrazinopyridine	510	–	2.978×10^3	–
122	4-(5-Chloro-2-pyridine)azo-1,3-diaminobenzene (5-Cl-PADAB)	570	H2SO4	6.38×10^4	–
123	Bromo-sulfonazo III	623.7	2.9	6.375×10^4	$2.13 \, \mu g/L$ (DL)

(Continued)

Table 1.1 *(Continued)* List of Spectrophotometric Reagents for Determination of Palladium

S. No.	Name of Analytical Reagent	Absorption Maximum (λ_{max}) (nm)	pH/Medium	Molar Absorptivity (L/mol cm)	Sandell's Sensitivity Detection Limit
124	2-(3,5-Dichloropyridylazo)-5-dimethylaminoaniline (3,5 dicl-PADMA) + CTAB		0.05–0.50 mol/L HCl		0.05 µg/L (DL)
125	3-Hydroxy-3-propyl-1-(4-carbamimidoyl sulfamoyl)phenyltriazene (CSPT)		1.8–2.2	8372	12.71 mg/cm
126	1-(2-Pyridylazo)-2-naphthol (PAN)	–	–	–	–
127	Allylthiourea				
128	2-(2-Quinolinylazo)-1,3-diaminobenzene (QADAB)	590	0.2–3.0 mol/L phosphoric acid	8.08×10^4	
129	p-Sulfobenzylidenerhodanine	530			
130	2,4-Dichlorophenylfluorone (DCIPF)	470	HAC-NaAc buffer	4.46×10^5	
131	2-(5-Carboxy-1,3,4-triazolylazo)-5-diethylamino aniline	585	HAC – NaAc 5.5	1.2×10^4	
132	4-(2-Pyridylazo)resorcinol and 1-(2-pyridylazo)-2-naphthol		HCl		
133	2-Hydroxy-3-nitroso-5-Me acetophenone oxime (HNMA)	430	0.0–4.0	15796	
134	4-(2-Thiazolylazo)-5-diethylamine-aminobenzene	590	4.0 Acetate buffer	1.08×10^5	
135	$NiCl_2$-NaH_2PO_2 system	395	NaOH	–	0.03 µg/L (DL)

(Continued)

Table 1.1 (Continued) List of Spectrophotometric Reagents for Determination of Palladium

S. No.	Name of Analytical Reagent	Absorption Maximum (λ_{max}) (nm)	pH/Medium	Molar Absorptivity (L/mol cm)	Sandell's Sensitivity Detection Limit
136	p-Iodochlorophosphonazo + Tween-60	638			
137	PVC functional membrane	520	3.4		0.015 µg/10 mL (DL)
138	4-Hydroxy-1-naphthalrhodanine	520	3.5		
139	2-(2-Quinolinylazo)-5-dimethylaminobenzoic acid (QADMAB)	630	3.5	–	–
140	2-(p-Carboxyphenylazo)benzothiazole (CPABT)	510	5.9–7.5	1.61×10^5	–
141	2-(2-Quinolylazo)-5-diethylaminobenzoic acid (QADEAB)	628	0.05–0.5	1.43×10^5	0.02 µg/L (DL)
142	4-(5-Chloro-2-pyridyl)-azo-1,3-diaminobenzene (5-Cl-PADAB)	570	3.0 mol/L H_3PO_4	6.39×10^4	
143	N-decylpyridine-4-amine from malonate media	410	0.025 M sodium malonate	1.9×10^5	0.065 µg/cm²
144	2-(5-Carboxy-1,3,4-triazolyl-5-dimethyl aminobenzoic acid) (CTZDBA)	548		9.32×10^4	
145	1-(2'-Benzothiazole)-3-(4'-carboxybenzene) triazene	490			
146	Tween 80-$(NH_4)_2SO_4$-PAR System	528	EDTA – NaOH	4.2302×10^4	0.026 µg/10 mL
147	1-(2-Pyridylazo)-2-naphthol (PAN) + Chloroform				

(Continued)

Table 1.1 (*Continued*) List of Spectrophotometric Reagents for Determination of Palladium

S. No.	Name of Analytical Reagent	Absorption Maximum (λ_{max}) (nm)	pH/Medium	Molar Absorptivity (L/mol cm)	Sandell's Sensitivity Detection Limit
148	(p-Methoxyphenyl)ethane-1,2-dione 1-oxime (HMPED)	420	1.1–3.2	3.087×10^4	3.45×10^{-3} µg/cm^2
149	4-(N'-4-imino-2-oxo-thiazolidine-5-ylidene)-hydrazinobenzoic acid (p-ITYBA)	450	7.0	4.30×10^3	0.23 µg/mL (DL)
150	2-(5-bromo-2-pyridylazo)-5-[N-n-propyl-N-(3-sulfopropyl)amino]aniline (5-Br-PSAA)	612			
151	N-amyl-N'-(sodium-p-aminobenzenesulfonate)thiourea (APT)	294.4	HACl – NaAc Buffer	3.52×10^5	
152	p-Dimethylaminobenzaldehyde-(p-anisyl)-thiosemicarbazone (DBATSC) + glyoxal (p-anisyl)thiosemicarbazone (GATSC)				
153	Dye (Yellow-6 Y6)	553	3.0–9.5 7.0	1.14×10^4	
154	2-Acetylthiopyran thiocyanate (1-thiopyran-2-yl)ethanone thiocyanate (ATPT)	520	6.0		

Table 1.2 List of Spectrophotometric Reagents for Determination of Platinum

S. No.	Name of Analytical Reagent	Absorption Maximum (λ_{max}) (nm)	pH/Medium	Molar Absorptivity (L/mol cm)	Sandell's Sensitivity Detection Limit
1	Carbon–polyurethane powder	–	4–5	–	2.4 µg/L
2	Leucoxylene cyanol FF (LXCFF)	620	1.0–2.5 3.0–4.5	5.1×10^4	0.0038 µg/cm^2
3	1,3-Dimethyl-2-thiourea and bromocresol green	413	Acetate buffer H$_2$SO$_4$ 3.2–4.2	–	–
4	Astrafloxin FF	–	–	$8.1–13.3 \times 10^4$	–
5	Piperonal thiosemicarbazone	360	0.008–0.32 M H$_2$SO$_4$	3.239×10^4	0.006 µg/cm^2
6	4-[N,N-(diethyl)amino]benzaldehyde thiosemicarbazone	405	–	1.755×10^4	0.0012 µg/cm^2
7	1-Phenyl-4-ethyl thiosemicarbazide	715	3	0.14×10^5	–
8	N-(3,5-Dimethylphenyl)-N'-(4-aminobenzenesulfonate) thiourea (DMMPT)	755	3.8	9.51×10^4	–
9	Phenanthrenequinonemonosemicarbazone	–	–	–	–
10	Anthranilic acid	375	3.0	3.41×10^4	5.6×10^{-3} µg/cm^2
11	SnCl$_2$ + HCl and cetylpyridinium chloride (PC) + Triton X-100	405	HCl	3.00×10^3	150 ng/mL (DL)
12	Isonitroso p-methyl acetophenone phenylhydrazone (HIMAPH)	465	5.0–7.2	10731.71	–

(Continued)

Table 1.2 (Continued) List of Spectrophotometric Reagents for Determination of Platinum

S. No.	Name of Analytical Reagent	Absorption Maximum (λ_{max}) (nm)	pH/Medium	Molar Absorptivity (L/mol cm)	Sandell's Sensitivity Detection Limit
13	Potassium bromate and 2-(4-chloro-2-phosphonophenyl)azo-7-(2,4,6-trichlorophenyl)azo-1,8-dihydroxy-3,6-naphthalene disulfonic acid	—	—	—	—
14	1-(2-Pyridylazo)-2-naphthol	690	2.5–6.5	1.6×10^5	0.0048 mg/cm^2
15	2-(5-Bromo-2-oxoindolin-3-ylidene) hydrazine carbothioamide (HBITSC)	505	3.6–6.0	8452.53	23.07 ng/cm^2
16	o-Methoxyphenylthiourea and iodide	362	Aqueous KI (0.1 mol/L)	1.25×10^4	0.016 µg/cm^2
17	o-Aminobenzylidenerhodanine (ABR)	525	HCl	7.7×10^4	—
18	p-Aminobenzylidenerhodanine (ABR)	525	HCl	5.98×10^4	—
19	2-Hydroxy-1-naphthalrhodanine	540	HCl	1.02×10^5	—
20	p-Sulfobenzylidenerhodanine (SBDR)	545	HCl	7.25×10^4	—
21	5-(H)-acidazorhodanine (HAR)	540	HCl	1.06×10^5	—
22	5-(2-Hydroxy-4-sulfo-5-chlorophenol-1-azo)thiorhodanine (HSCT)	535	HCl	6.24×10^4	—
23	2-Hydroxy-5-sulfobenzenediazoaminoazo benzene (HSDAA)	434	11.7	1.59×10^5	1.14×10^{-6} g/L (DL)
24	Isochromatic ion pair between tetrabromofluorescein and rhodamine 6G	530	5.5 HAC–NaAC	9.94×10^5	—

(Continued)

Table 1.2 (Continued) List of Spectrophotometric Reagents for Determination of Platinum

S. No.	Name of Analytical Reagent	Absorption Maximum (λ_{max}) (nm)	pH/Medium	Molar Absorptivity (L/mol cm)	Sandell's Sensitivity Detection Limit
25	2-(5-Iodine-2-pyridylazo)-5-dimethylaminoaniline (5-I-PADMA)	625	3.7–5.6	4.55×10^4	–
26	Polyamide resin	–	HCl (1–3)	–	–
27	3,4-Diaminobenzoic acid (DBA) + Molten naphthalene	715	10.0–12.5	1.2×10^6	$0.0041 \ mg/cm^2$
28	2-(p-Carboxyphenylazo)benzothiazole (CPABT) + Tween-80	508	HCl NaAC (3.8–5.5)	2.29×10^5	–
29	4,5-Dibromo-2-nitrophenylfluorone (DBON-PF)	476	HAC – NaAC 3.6	1.88×10^6	$1.06 \times 10^{-6} \ g/L$ (DL)
30	N-(3,5-dihydroxyphenyl)-N'-(4-aminobenzenesulfonate)thiourea (DHPABT)	760	0.05–0.5 M HCl	1.01×10^5	$0.02 \ \mu g/L$ (DL)
31	5-(5-Iodo-2-pyridylazo)-2,4-diaminotoluene (5-I-PADAT)	595	5.5	3.2×10^4	–
32	p-Rhodanineazobenzoic acid Na dodecylbenzenesulfonate (SDBS) (RABA + SDBS)	520	KOH C_6H_{42} HK buffer solution (4.0–6.2)	2.49×10^5	–
33	4-(2'-Furalidineimino)-3-methyl-5-mercapto-1,2,4-triazole in n-butanol (FIMMT)	510	5.4 (HAC/NaAC) 0.2 M	1.1686×10^4	$0.017 \ \mu g/cm^2$

an effort to continue this to provide researchers a single-source book for spectrophotometric determination of palladium and platinum developed during the last 20 years.

To apprise readers with the recently used spectrophotometric reagents, Tables 1.1 and 1.2 give names of analytical reagents for Pd and Pt and other brief details such as λ_{max}, pH, molar absorptivity, Sandell's sensitivity or detection limit, etc. This can be a very useful comparative review for any analyst interested in choosing a particular reagent or method.

References

1. Goswami, A.K. and Agarwal, S. *Spectrophotometric Determination of Copper and Iron, Reagents and Methods*. De Gruyter, GmbH, Berlin/Boston, 2021.
2. Goswami, A.K. and Kostova, I. *Medicinal and Biological Inorganic Chemistry*. De Gruyter, GmbH, Berlin/Boston, 2022.
3. Rossotti, J.C. and Rossotti, H. *The Determination of Stability Constants*. McGraw Hill, New York, 1961.
4. Dalawat, D.S., Chauhan, R.S. and Goswmi, A.K. Review of spectrometric methods for the determination of Zirconium. *Rev. Anal. Chem.*, 2005, 24(2), 75–102.
5. Khanam, R., Singh, R., Mehta, A., Dashora, R., Chauhan, R.S. and Goswami, A.K. Review of spectrophotometric methods for the determination of Nickel. *Rev. Anal. Chem.*, 2005, 24(3), 149–245.
6. Singh, K., Chauhan, R.S. and Goswami, A.K. A review of reagents for spectrophotometric determination of Lead(II). *Main Group Metal Chem.*, 2005, 28(3), 119–149.
7. Upadhyay, M., Chauhan, R.S. and Goswami, A.K. A review of spectrophotometric reagents for Cadmium determination. *Main Group Metal Chem.*, 2005, 28(6), 301–357.
8. Khan, S., Dashora, R., Goswmi, A.K. and Purohit, D.N. Review of spectrophotometric methods for determination of iron. *Rev. Anal Chem.*, 2003, 22(1), 73–80.
9. Ram, G., Chauhan, R.S., Goswami, A.K. and Purohit, D.N. Review of spectrophotometric methods for determination of Cobalt(II). *Rev. Anal. Chem.*, 2003, 22(4), 255–317.
10. Gorji, D.K., Chauhan, R.S., Goswami, A.K. and Purohit, D.N. Hydroxytriazenes – a review. *Rev. Anal. Chem.*, 1998, 17(4), 223–234.
11. Ressalan, S., Chauhan, R.S., Goswami, A.K. and Purohit, D.N. Review of spectrophotometric methods for determination of Chromium. *Rev. Anal. Chem.*, 1997, 16(2), 69–171.
12. Rezaei, B., Zabeen, R., Goswami, A.K. and Purohit, D.N. Review of spectrophotometric methods for determination of Vanadium. *Rev. Anal. Chem.*, 1993, 12(1–2), 1–200.
13. Zabeen, R., Rezaei, B. and Purohit, D.N. Spectrophotometric methods for determination of Palladium: a review. *Rev. Anal Chem.*, 1991, 10(2), 115–358.

section A

---

*Spectrophotometric
Determination of Palladium –
Reagents and Methods*

chapter two

Analytical Reagents Having Oxygen (O) as Donor Atom

Analytical reagents that have oxygen atom as donor are included in this chapter. The methods with very precise details are described herein. An attempt has been made to give spectrophotometric determination parameters like wavelength at which the complex absorbs pH and medium, working wavelength or λ_{max}, molar absorptivity, Sandell's sensitivity and detection limit of reagent. Thus, the method with a particular reagent can be applied by anyone who wants to determine palladium with his/her choice of conditions and sample.

2.1 Naphthols

2.1.1 1-Nitroso-2-Naphthol

Hussain et al. [1] have described a method for the determination of palladium contents using 1-nitroso-2-naphthol in alkaline medium. The Pd contents using spectrophotometric method are based on application to Rasagiline mesylate sample in which palladium complex with the named reagent is extracted with toluene. The complex absorbs at 425 nm and has a detection limit (LOD) of 0.06 ppm. The developed method was subject to validation for specificity, precision, limit of detection, linearity, limit of quantitation, accuracy, robustness and solution stability as claimed by the authors.

2.1.2 1-(5-Benzylthiazol-2yl) Azonaphthalen-2-ol (BnTAN)

Tupys and Tymoshuk [2] have introduced a simple, selective and inexpensive extraction spectrophotometric method for determination of Pd(II) using BnTAN, an azo dye. It is mentioned that spectrophotometry is a popular method in the laboratories of developing countries and extraction spectrophotometry is extremely suitable owing to its advantages like simple and easy handling and even low cost of instrumentation. The Pd(II) ion interacts with BnTAN to form a chelate. The complex is then extracted into toluene and chloroform from the alkaline and acid aqueous solutions,

DOI: 10.1201/9781003276418-4

respectively. The extracts thus obtained show absorbance maxima at 600 and 684 nm, respectively. It is mentioned that Beer's law is obeyed in the range of 0.261–0.850 μg/mL of Pd(II), with a molar absorptivity of the complex of 6.70×10^3 L/mol cm. The method is successfully applied in the determination of Pd(II) in electronic devices as mentioned by the authors.

2.1.3   5-Hydroxyimino-4-Imino-Thiazolidin-2-One (HITO)

Lozynska et al. [3] have reported spectrophotometric determination of Pd(II) using HITO in acetate buffer and sodium chloride medium of pH 3.0. The method as stated by the authors has been successfully applied to measure palladium in two intermetallides, viz. Yb_{40} Pd_{38} Sn_{22} and Yb_{40} Pd_{40} Ga_{20}. The applicable range of the determined concentrations of metal lies between 6.0×10^{-6} and 6.0×10^{-5} M which is good. Further, the method is characterized by its excellent selectivity as well as no interference of most of the contaminant ions. It is advantageous over other methods due to its simplicity, rapidity, no use of any masking agent or previous separation of the analyte from other components of intermetallides. As mentioned by the authors the results correlate with the nominal content of palladium in the alloys and are confirmed by voltammetry.

2.1.4   3-Hydroxy-2-(2'-Thienyl)-4-H-Chromen-4-One

A rapid, simple, selective and sensitive method has been reported by Naini et al. [4] for determining trace palladium. The named reagent reacts with palladium in alkaline medium, pH 8.5–9.2 to form a yellow-coloured complex, extracted into chloroform. At an absorbance maximum of 455 nm it is measured in the Beer's law range of 0.01–0.1 μg Pd/mL. There is no interference of large numbers of cations, whereas anions/complexing agents like nitrite, thiourea and ascorbic acid show interference. As reported the molar absorptivity and Sandell's sensitivity are 3.301×10^4 L/mol cm and .0032 μg Pd/cm, respectively. The metal to ligand ratio is reported as 1:1 with quite reproducible results; validity of the method has a standard deviation of 0.0021% for ten replications as reported by the authors.

2.1.5   Dibromo-O-Nitro-Phenylfluorone (DBON-PF)

The chromogenic reaction between Pd(II) and DBON-PF has been studied and reported by Li et al. [5]. It is described that the reagent forms a 1:2 stable complex in HCl solution at pH 2.2 in the presence of microemulsion (m(CPB):m (n-BuOH):m (n-heptane):m = 1:1:1:97). The λ_{max} is 474 nm and Beer's law range is 0.1–1.0 μg/mL of Pd(II). The molar absorptivity is found to be 1.07×10^5 L/mol cm. It is stated that the method is used for determination of Pd in Pd catalysts with consistency of the results with AAS.

2.1.6 Pyridyl-Azo-Naphthol (PAN)

A new combined method including fibreoptic linear array detection spectrophotometry (FO-LADS) and liquid-liquid micro-extraction (DLLME) using a cylindrical micro-cell has been developed by Shokoufi et al. [6] for detection of various species. The method has been successfully used as a separation cum detection method in which an appropriate mixture of ethanol (dispersing solvent) and 1,2-dichlorobenzene (extracting solvent) is injected into the water sample which contains palladium and cobalt after complexing with 1-(2-pyridylazo)-2-naphthol (PAN). Once the phase separation is complete 50 µL of the sedimented phase which contains Pd is determined by FO-LADS. Both ordinary and first-derivative spectra are obtained to simultaneously determine Pd and Cobalt. The proposed method is comparable with other methods as described by the authors.

2.1.7 4-(2-Pyridylazo)-Resorcinol (PAR)

As reported by Dong and Gai [7], a red complex is formed by Pd with PAR at 90°, in the pH range of 9.0–11.0 and complex is extracted into molten naphthalene. At λ_{max} value of 520 nm the organic phase dissolved in $CHCl_3$ can be detected against reagent blank. Further, Beer's law is obeyed in the concentration range of 0.1–1.2 µg/mL. Reported values of molar absorptivity and Sandell's sensitivity for the said complex are 8.0×105 L/mol cm and 0.49 µg. Most of the ions do not interfere yet Co(II), Fe(II) and Bi(III) are effectively screened using EDTA as informed.

2.1.8 2-(2-Quinolinylazo)-5-Dimethyl Aminobenzoic Acid (QADMAB)

Hu et al. [8] have reported synthesis and colour reaction of 2-(2-quinolinylazo)-5-diaminobenzoic acid with Pd(II). The authors report the reagent as extractant for the extractive spectrophotometric determination of Pd(II). The reagent gives a yellow complex which is extracted into Butan-1 at pH 4.0. The measurement is made at 410 nm, giving molar absorptivity of the complex as 0.672×10^2 L/mol cm, whereas Sandell's sensitivity is found to be 0.06729 µg/mL. It is mentioned that the method can be applied for Pd catalysts in any binary mixture.

2.1.9   α-Benzyl Monoxide-Molten Benzophenone

Eskandari et al. [9] have reported a selective and sensitive method for preconcentration of Pd(II) from H_2SO_4 solution applying α-benzyl monoxide-molten benzophenone as reagent. It is described that Pd can be extracted quantitatively into molten benzophenone which can be detected spectrophotometrically at λ_{max} of 434 nm. The Beer's law range is 0.03–0.60 µg/mL of Pd (1.5–3.0 µg) with a detection limit of 3 mg/mL. It is mentioned that no serious interference by other ions is observed. It is further mentioned that the method being precise, accurate and selective can be successfully applied to determination of Pd in synthetic mixtures.

2.1.10   1-(2-Pyridylazo)-2-Naphthol (PAN)

Patil et al. [10] have proposed an extraction spectrophotometric method for the determination of Pd(II). Pd(II) can be extracted from salicylate solution using triphenylphosphine sulphide as an extractant. Pd(II) in the organic phase then is determined with 1-(2-pyridylazo)-2-naphthol. It is stated that the method can be successfully applied for separation of Pd(II) from Pt(IV), Rh(III), Ir(III), Ru(IV), Os(VIII), Hg(II), Cu(II), Co(II) and Ni(II) and even determination of Pd(II) in Pd catalysts.

2.1.11   2-Arylthio-p-Nitro-Acetophenone

Gojare et al. [11] have brought forward a very simple, rapid and selective method for the spectrophotometric determination of Pd(II) using extraction of the complex in $CHCl_3$. The pH mentioned is 7–8 M in HOAC medium. A yellow-coloured Pd(II)-2-arylthio-p-nitro-acetophenone complex has a λ_{max} of 430 nm with molar absorptivity of 1612 L/mol cm. It is described that common anions and cations do not interfere in the determination of Pd(II). It is mentioned by the authors that the entire process of extraction and determination takes 15 minutes.

2.1.12   Pyrogallol Red (PGR) with H_2O_2

A kinetic spectrophotometric method for Pd(II) determination is described by Ensafi and Keyvanfard [12]. This is based on the catalytic action of Pd(II) on the oxidation of pyrogallol red (PGR) with H_2O_2 at pH 9.7. The reaction can be monitored spectrophotometrically by measurement of decrease in the absorbance of PGR at 540 nm within the first 4.5 minutes from initiation of this reaction. The calibration is reported to be linear in the range of 0.02–1.00 µg/mL of Pd(II). The interference of 740 species also has been studied for selectivity. Authors state that the proposed method is applicable to Pd(II) determination in catalytic material.

2.1.13   Pyridylazo Resorcinol

Chakkar and Kakkar[13] proposed a spectrophotometric determination of Pd(II) using pyridylazo resorcinol as a ligand. The complex is extracted into chloroform in the presence of diethylamine. λ_{max} is observed at 535 nm, with a Sandell's sensitivity of the complex as 0.0034 mg Pd/cm². It is further mentioned that large numbers of platinum group elements do not interfere. The method can be applied to analysis of various samples with satisfactory and reproducible results with a standard deviation of ±0.002.

References

1. Hussain, S., Gosar, A. and Shaikh, T. Development and validation of UV-spectrophotometric method for determination of palladium content in Rasagiline Mesylate. *Br. J. Pharm. Res.*, 2017, 15(4), 1–7.
2. Tupys, A. and Tymoshuk, O. 1-(5-benzylthiazolyl-2-yl) azonaphthalen-2-ol-a new reagent for the determination of Pd(II). *Acta Chim. Slov.*, 2015, 8(1), 59–64.
3. Lozynska, L., Tymoshuk, O. and Rydchuk, P. Spectrophotometric method for palladium determination using 5-hydroxyimino-4-imino-1,3-thiazolidin-2-one and application to analysis of intermetallides. *Chem. Metals Alloys*, 2014, 7, 119–122.
4. Naini, S., Agnihotri, N. and Kakkar, L.R. Extractive spectrophotometric determination of palladium with 3-hydroxy-2-(2'-thienyl)-4H-chromen-4-one in alkaline medium. *Int. J. Sci. Eng. Res.*, 2013, 4(9).
5. Li, Y., Xu, X., Ma, W., Xu, G. and Fang, G. Spectrophotometric determination of palladium with dibromo-o-nitro-phenyl fluorone in microemulsion medium. *Yejin Fenxi*, 2008, 28(5), 54–56.
6. Shokoufi, N., Shemirani, F. and Assadi, Y. Fiber optic-linear array detection spectrophotometry in combination with dispersive liquid-liquid microextraction for simultaneous preconcentration and determination of palladium and cobalt. *Anal. Chim. Acta*, 2007, 597, 349–356.
7. Dong, Y. and Gai, K. Spectrophotometric determination of palladium after solid-liquid extraction with 4-(2-pyridylazo) resorcinol at 90°C. *Bull. Korean Chem. Soc.*, 2005, 26(6), 143–946.
8. Hu, Q., Zhu, L., Chen, J., Yang, G. and Yin, J. Spectrophotometric determination of palladium with 2-(2-quinolinylazo)-5-dimethylaminobenzoic acid. *Yankuangceshi*, 2004, 23(2), 117–120.
9. Eskandari, H., Ghaziaskar, H.S. and Ensafi, A.A. Solid-liquid separation after liquid-liquid extraction using α-benzyl monoxime-molten benzophenone for preconcentration and selective spectrophotometric determination of palladium. *Anal. Lett.*, 2001, 34(14), 2535–2546.
10. Patil, N.N. and Shinde, V.M. Extraction and spectrophotometric determination of palladium (II) with triphenylphosphine sulfide. *Ind. J. Chem.*, 1997, 34A(4), 347–348.

11. Gojare, P.T., Gaikwad, S.H. and Anuse, M.A. Selective solvent extraction and spectrophotometric determination of palladium (II) with 2-arylthio-p-nitro-acetophenone. *Res. J. Chem. Environ.*, 2001, 5(3), 51–56.
12. Ensafi, A.A. and Keyvanfard, M. Kinetic spectrophotometric determination of palladium in hydrogenation catalyst by its catalytic effect on the oxidation of pyrogallol red by hydrogen peroxide. *Spectro Chim. Acta Part A: Mol. Biomol. Spectr.*, 2002, 58A(8), 1567–1572.
13. Chakkar, A.K. and Kakkar, L.R. Spectrophotometric determination of palladium with 4-(2-pyridylazo) resorcinol. *Fresenius J. Anal. Chem.*, 1991, 340(1), 19–21.

chapter three

Analytical Reagents Having Nitrogen (N) as Donor Atom

3.1 2-(5-Bromo-2-Pyridylazo)-5-[N-n-Propyl-N-(3-Sulfopropyl)Amino] Aniline

A method proposed by Quezada et al. [1] describes simultaneous injection effective mixing flow analysis (SIEMA) and determination of palladium in dental alloys and hydrogenation catalysts with high accuracy. The method is based on complex formation of Pd with 2-(5-bromo-2-pyridylazo)-5-[N-n-propyl-N-(3-sulfopropyl)amino] aniline (5-Br-PSAA) which is a blue complex with λ_{max} at 612 nm. Advantages mentioned by the authors include parameters like reagent consumption, waste volume and time of analysis compared to conventional FIA method. It is also mentioned that this can be used for other chemical analyses.

3.2 L-Cystine

Chandra Shekhara et al. [2] have proposed a simple, rapid and reliable spectrophotometric method of Pd(II) determination using L-cystine as a ligand in alloy composition. Advantages of the method described by the authors include extraction, heating and being interference free, and the reagent is non-toxic and easily available. The reagent forms a yellow complex absorbing at 369 nm, following the Beer's law range of 2.12–16.9 µg/L of Pd(II). The molar absorptivity and Sandell's sensitivity value mentioned are 2.65×10^4 L/mol cm and 7.89×10^{-4} µg/cm^2 respectively. Interference studies on different ions as well as application of the method to synthetic alloy sample have also been mentioned.

3.3 2-Hydrazinopyridine

Soliman et al. [3] have brought forward a rapid, simple, selective and validated spectrophotometric method of Pd(II) determination using 2-hydrazinopyridine as a ligand. The complex so formed can be quantitatively measured at 510 nm under optimized conditions. As mentioned, the stoichiometry and stability constant of a purple-coloured complex so formed

DOI: 10.1201/9781003276418-5

are studied spectrophotometrically at 25°C using two methods, viz. Job's continuous variation and mole ratio method. A 1:1 (M:L) ratio and 1.06–9.00 µg/mL Beer's law range are reported, with a molar absorptivity of the complex at 2.978×10^3 L/mol cm. Recovery percent of (96.61–102.58) and RSD% (0.04–0.41) are also reported. The advantage of non-interference of different ions in the Pd(II) determination is also reported by the authors.

3.4   2-Hydroxy-5-Methylacetophenone Isonicotionoylhydrazone (HMAINH)

The mentioned reagent has been used by Pethe et al. [4] for spectrophotometric determination of Pd(II). The method being a simple, rapid and selective one involves a 1:1 yellow complex with Pd(II) which absorbs at 385 nm against reagent blank. The complex so formed is extracted at 0.010–0.015 M H_2SO_4 into chloroform. The entire concentration range between 2.0 and 9.0 ppm of Pd(II) obeys Beer's law. Further the molar absorptivity and Sandell's sensitivity values as reported are 5.320×10^3 dm³/mol cm and 0.02 µg/cm² respectively with standard deviation of 0.8755. Various ions as mentioned do not interfere in the determination of Pd(II) in the extraction method.

3.5   Isonitroso-p-Me.acetophenonephenyl Hydrazone (HIMAPH)

Kumar et al. [5] have reported an extractive spectrophotometric method using HIMAPH. It is mentioned that Pd(II) can be quantitatively extracted up to 99.79 using HIMAPH from an aqueous solution of pH 0.0–5.0 from 0.1 to 1 M solution of acetic acid and mineral acids in the presence of 1 mL of 2 M NaAC solution, which is followed by digestion on boiling water bath for 4–5 minutes. The toluene extract shows λ_{max} at 470 nm with 0.1–10 µ/mL concentration range of Beer's law. Two methods, viz. Job's method and mole ratio method, have been used for the determination of complex composition as 1:2 (Pd:L). Even the interference of diverse ions is not reported and the method can be successfully applied for Pd(II) determination in alloys as well as catalyst samples.

3.6   1-(2-Naphthalene)-3-(2-Thiazole)-Triazene (NTT)

A new reagent NTT has been synthesized, characterized and applied for the spectrophotometric determination of Pd(II) by Long et al. [6]. It is mentioned that NTT as a chromogenic reagent increases absorbance in the concentration range of Pd(II) at 01.0–1.75 µg/mL. Absorption coefficient is reported as 4.07×10^4 L/mL/cm, and the method has been successfully and satisfactorily used for Pd determination in the catalyst samples.

3.7 p-Antipyrinylazo Benzoic Acid (AABA)

Huang et al. [7] have reported a new reagent p-antipyrinylazo benzoic acid (AABA) for the spectrophotometric determination of Pd(II). It is described that in the presence of emulsifier OP in HAC-NaAC buffer of pH 3.5–5.5 the reagent reacts with Pd(II) to give a stable 2:1 complex. The coloured complex is extracted by C18 cartridge and eluted with ethanol. The complex absorbs at 452 nm and is determined using spectrophotometry. The Beer's law range reported is 0.1–1.5 µg/mL for Pd with a molar absorptivity of 1.22×10^5 L/mol cm. It is mentioned that the method can be applied to determine trace Pd in Pt/Pd catalysts and results are comparable to those obtained using AAS.

3.8 3,4-Dihydroxybenzaldehydeisonicotinoyl-hydrazone (3,4-DHBINH)

A very simple, rapid, sensitive and inexpensive method proposed by Narayana et al. [8] uses the mentioned reagent 3,4-DHBINH for spectrophotometric determination of Pd(II). It is reported that a yellow complex of Pd(II) is formed at pH 3.0 in 1:1 (M:L) composition. The maximum absorbance of the complex is at 380 nm in the Beer's law range of 0.5–20.0 ppm. The reported molar absorptivity and Sandell's sensitivity are 0.53×10^4 L/mol cm and 0.02 µg/cm², whereas the detection limit is 0.0948 µg/mL. The correlation coefficient for the complex is 1.08 and 0.04. Interference of various cations and anions is none, making the method more applicable. It is further reported that the developed method can be successfully applied to Pd(II) analysis in spiked samples. The validity of the method is tested using AAS and the method has been used for the determination of Pd(II) in water as well as synthetic mixtures.

3.9 N,N,N',N'-Tetra(2-ethylhexyl) Thiodiglycomide T(2EH) TDGA

A precise, sensitive and selective method of Pd(II) determination has been reported by Ruhela et al. [9] using T(2EH) TDGA as an extractant. It is reported that Pd(II) forms a yellow-coloured complex with this ligand which absorbs at 300 nm. The Beer's law applicability is in the range of concentration between 1.0 and 15.0 µg/mL of Pd, and the molar absorptivity is reported as 1.29×10^5/M/cm. It is further mentioned that relative standard deviation of the method is <0.5% and the method is precise. The method is successfully applied for Pd determination in simulated high-level liquid waste (SHLW) solution.

3.10   2-Hydroxy-1-naphthalene Carboxaldehyde Hydrazine Carboxamide

Lokhande and Saini [10] have reported a spectrophotometric determination method for Pd(11) using 2-hydroxy-1-naphthalene carboxaldehyde hydrazine carboxamide as reagent. It forms an orange complex with Pd(II) which can be quantitatively determined after extraction in EtoA at pH 4.2. The complex extracted absorbs at 410 nm, for a Beer's law concentration range of 0.55 g/mL to 2.50 μg/mL of Pd. Reported molar absorptivity and Sandell's sensitivity values are 0.4×10^3 L/mol cm and 1.50×10^{-5} μg/cm^2, respectively. The reported stoichiometry of the complex is 1:2, and the method has been successfully applied to determine Pd in Ag alloys, Pd complexes and Pd catalysts.

3.11   1-(4-Antipyrine)-3-(2-thiazolyl) triazene (ATTA)

Bin et al. [11] have brought forward synthesis and application of 1-(4-antipyrine)-3-(2-thiazolyl)triazene (ATTA) for Pd(II) determination. The duly characterized reagent is reported to be a highly selective as well as sensitive reagent for Pd(II) determination. It is mentioned that a brown-coloured complex of Pd(II), ATTA absorbs at λ_{max} of 480 nm in the presence of OP and $Na_2B_4O_7$-NaOH buffer at pH 9.5. The linear range (Beer's law) is 0.01–12.60 μg/mL of Pd(II) with a molar absorptivity of 5.33×10^4 L/mol cm.

3.12   N-Bulyl-N'-(sodium p-amino-benzenesulfonate) Thiourea (BPT)

Hou et al. [12] have reported a highly sensitive reagent N-bulyl-N'-(sodium p-amino-benzenesulfonate) thiourea (BPT) for spectrophotometric determination of Pd. It is reported that in the presence of cetyltrimethylammonium bromide (CTMAB), Pd reacts with BPT to give a stable yellow complex in HOAC–NaOAC buffer solution of pH (4.6–5.6). The yellow complex absorbs at λ_{max} 294.4, and it has a molar absorptivity value of 2.46×10^5 L/mol cm. It is reported to be a rapid, simple and sensitive one for the trace amounts of Pd in minerals and catalyst samples.

3.13   Methdilazine Hydrochloride

Meiwaukl et al. [13] have reported a new reagent methdilazine hydrochloride for determination of Pd(II) using spectrophotometry. It is mentioned that the reagent forms a violet-coloured complex instantaneously with Pd(II) which has λ_{max} at 555 nm in buffer medium. The molar absorptivity of the complex is 1.66×10^4 L/mol cm with Sandell's sensitivity of 20.3 ng/cm^2. A 1:1 composition of the complex has been determined using Job's method and mole ratio method. It is mentioned that the Beer's law range is between 0.3 and 12.5 ppm of Pd(II), with no interference of foreign ions. The method can be applied successfully for the determination of Pd(II) in alloys and minerals.

3.14   2-Hydroxy-3-carboxy-5-sulfophenyldiazoaminoazobenzene (HCSDAA)

A new triazene 2-hydroxy-3-carboxy-5-sulfophenyldiazoaminoazobenzene (HCSDAA) has been reported by Zheng et al. [14]. It is reported that Pd(II) forms a 1:2 (M:R) complex with HCSDAA in the pH range of 9.3–10.6 in a buffer medium of NaHCO$_3$ in the presence of Triton N-101. Further, as reported the red complex absorbs at 530 nm in the linear range of 0–0.7 μg/mL. The masking agents K-Na tartarate and NaF are also used. A 1.05×10^5 of molar absorptivity is reported and the method has been successfully applied to Pd determination in source noble metal alloys after separation using dimethylglyoxime CHCl$_3$ extraction with an RSD of 2.0% with recovery value of 93.8%–99.3%.

3.15   3-Hydroxy-3-phenyl-1-p-carboxyphenyltriazene

Zabeen et al. [15] have reported a new triazene as chromogenic reagent for Pd(II) determination at room temperature. It is described that Pd(II) forms an instantaneous dark yellow complex with the reagent, which has a 1:2 (Pd:R) composition. The absorbance maxima at which the complex can be determined is reported to be 430 nm, with an optimum pH range of 3.0–3.5. Advantageously the colour is stable for 24 hours and the molar absorptivity and Sandell's sensitivity values of 14,000 L/mol cm and 7.60 ng/cm^2, respectively, are reported. RSD for ten determinations for 2.66 ppm of Pd has been found as 0.015 ppm (0.56%). Interference of foreign ions also has been worked out as reported.

References

1. Quezada, A.A., Noguchi, D., Murakami, H., Teshima, N. and Sakai, T., Simultaneous injection effective mixing flow analysis system for spectrophotometric determination of palladium in dental alloys and catalyst. *J. Flow Injection Anal.*, 2015, 32(1), 13–17.

2. Chandra Shekhara, K.G., Gopal Krishna, B.N. and Nagaraj, P. Facile and direct spectrophotometric determination of palladium (II) with L-cystine. *Int. J. Chem. Stud.*, 2015, 2(6), 1–4.

3. Soliman, A.A., Majeed, S.R. and Altaby, F.A. Spectrophotometric determination of palladium using 2-hydrazinopyridine. *Eur. J. Chem.*, 2014, 5(1), 150–154.

4. Pethe, G.B., Bhadange, S.G., Joshi, M.D. and Aswar, A.S. Extractive spectrophotometric determination of palladium (II) with isonitroso-p-methylacetophenoneisonicotinoylhydrazone (HMAINH). *Adv. Appl. Sci. Res.*, 2010, 1(2), 58–64.

5. Kumar, A., Gupta, S. and Barhate, V.D. Extraction and spectrophotometric determination of Pd(II) with isonitroso-p-Me-acetophenonephenyl hydrazone (HIMAPH). *Orient. J. Chem.*, 2010, 22(10), 7551–7556.

6. Long, W., Hong, T., Ding, Z. and Cao, Q. Studies on the synthesis of 1-(2-naphthalene)-3-(2-thiazole)-triazene and its application for the spectrophotometric determination of Pd(II). *Guijinshu*, 2008, 29(3), 33–36.

7. Huang, Z., Huang, F. and Xie, Q. Study on solid phase extraction and spectrophotometric determination of palladium with p-antipyrinyl azo benzoic acid. *Yejin Fenxi*, 2007, 27(11), 17–20.

8. Narayana, S.L., Ramachandraiah, C., Reddy, A.V., Lee, D and Shim, J. Determination of traces of Pd(II) in spiked samples by using 3,4-dihydroxy-benzal-dehydeisonicotinoylhydrazone as a chelating agent with UV visible spectrophotometer. *E- J. Chem.*, 2011, 8(1), 217–225.

9. Ruhela, R., Sharma, J.N., Tomar, B.S., Hubli, R.C. and Suri, A.K. Extractive spectrophotometric determination of palladium with N, N, N', N'-tetra (2-ethylhexyl)-Thioglycolamide T(2EH) TDGA, *Talanta*, 2011, 85(2), 1217–1220.

10. Lokhande, R.S. and Saini, N. Extractive spectrophotometric determination of palladium (II) using 2-hydroxy-naphthalenecarboxyaldehyde hydrazine carboxamide as an analytical reagent. *Asian J. Chem.*, 2007, 19(1), 159–164.

11. Bin, G.Y., Cao, Q., Li, C. and Wang, J. Studies on synthesis of 1-(4-antiprine)-3-(2-thiazolyl) triazene and its application for spectrophotometric determination of palladium (II). *Fenxi Shiyanshi*, 2006, 25(5), 19–22.

12. Hou, F., Ma, D., Li, J. and Wang, Y. Spectrophotometric determination of palladium by the new chromogenic reagent N-butyl-N'-(sodium-p-aminobenzenesulfonate) thiourea. *Anal. Lett.*, 1998, 31(11), 1929–1936.

13. Meiwaukl, M.B. and Seetharamappa, J. Spectrophotometric determination of palladium (II) using methdilazine hydrochloride. *J. Saudi Chem. Soc.*, 2000, 4(2), 165–168.

14. Zheng, G., Guo, Z., Shao, Y. and Zhang, S. Spectrophotometric determination of palladium with triazene reagent (HCSDAA). *Lihua Jianyan, Huaue Fence*, 1999, 35(1), 14–15.

15. Zabeen, R., Goswami, A.K. and Purohit, D.N. 3-hydroxy-3-phenyl-1-p-carboxy phenyltriazene a new reagent for spectrophotometric determination of palladium. *Chim. Acta Turc.*, 1994, 22(2), 221–224.

chapter four

Analytical Reagents Having (N) and (O) as Donor Atoms

4.1   4-(4'-Fluorobenzylideneimino)-3-methyl-5-mercapto-1,2,4-triazole (FBIMMT)

Shaikh et al. [1] have proposed a simple, rapid and selective method for the determination of Pd(II) using FBIMMT as reagent. It is reported that the mentioned reagent instantaneously forms a 1:1 yellow-coloured complex which is extracted in chloroform in HCl medium. The extracted complex shows a maxima at 390 nm. The Beer's law concentration range up to 17.5 µg/mL is reported with 5.0–17.5 µg/mL, as calculated using Ringbom's plot. The parameters such as optimum extraction conditions, reagent concentration, acidity, solvents and interferences of various anions and cations as well as even shaking time are included in this chapter. The molar absorptivity and Sandell's sensitivity are 5.404×10^3 L/mol cm and 0.0196 µg/cm^2, respectively. It is further mentioned that the selectivity of method may be enhanced using masking agents. Thus, as reported the method has been successfully applied for the separation and determination of palladium (II) from binary mixtures, multicomponent synthetic mixtures, synthetic mixtures as well as alloys and catalysts.

4.2   1-(2-Benzothiozolyazo)-2-hydroxy-3-naphthoic acid (BTAHN)

Hassan and Amin [2] have reported a solid-phase extraction-based sensitive method for Pd(II) determination using BTAHN as reagent. The method describes details of extraction of BTAHN-Pd(II) chelate with a reversed-phase polymer-based C$_{18}$ cartridge, a fast procedure to determine Pd. In 5.0 HCl solution and in the presence of CTAB, BTAHN reacts with Pd(II) to form a deep violet-coloured complex. It has a stoichiometry of 2:1 [BTAHN:Pd (II)]. The complex is enriched by CPE with reverse-phased C$_{18}$ cartridge. Use of small amount of isopentyl alcohol enriches elution by a factor of 500. The spectrophotometric parameters as mentioned are the Beer's law concentration range of 0.02–0.85 µg/mL, molar absorptivity of 2.61×10^5 L/mol cm and λ_{max} of 699 nm. The detection limits as well as

DOI: 10.1201/9781003276418-6

quantification limits are reported as 6.3 and 19.65 ng, respectively. RSD values of 1.06% for 11 replications are further reported. It is claimed by the authors that the method can be successfully applied to the determination of Pd in environmental samples achieving excellent results.

4.3   2,4-Dihydroxy-5-iodo-α-phenyl acetophenone oxime [DHI-α-PAO]

A gravimetric and spectrophotometric method for Pd(II) determination using [DHI-α-PAO] is reported by Patel [3]. At pH 2.0, a 1:2 complex determined by Job's continuous variation as well as the Yoe and Jones method absorbs at 420 nm. The molar absorptivity and Sandell's sensitivity values of 4.23×10^2 L/mol cm and 0.2516 µg/cm^2, respectively, are mentioned. As reported the stability constant worked out spectrophotometrically has been found to be 4.30×10^9, whereas Gibb's free energy change is 13.226 kcal/mol. Linearity as defined by Beer's law is up to 74.49 ppm of Pd(II). The energy of activation for the decomposition of complex as revealed from TGA has been calculated using the Broido method as reported and found to be 14.76 and 8.443 kcal/mol for steps I and II, respectively. It is mentioned that the method can be applied for Pd determination in palladised carbon.

4.4   α-Furildioxime

A simple, rapid and efficient dispersive liquid–liquid microextraction-based UV–vis spectrophotometric method is proposed by Kozani et al. [4] for preconcentration and determination of Pd ions in water samples. It is described that Pd ions react with α-furildioxime (ligand) to give a hydrophobic chelate. The spectrophotometric determination parameters have been worked out and optimized. Effects of these on extraction efficiency pH, ligand concentration, volume of extraction and dispersive solvents, extraction time and salt concentration are also detailed out. It is mentioned that under optimized conditions the method showed an enrichment factor (C_{org}/C_{aq}) of 25 for recovery of more than 98%, which was a very short extraction time. The Beer's law linearity range is 10–200 µg/L with detection limit as 1.1 µg/L. The RSD for 100 µg/L concentration of Pd is 2.3% (n=10). The developed method has successfully been applied for the extraction and determination of Pd in tap water, river water, minerals and sea water samples. As claimed by the authors, the results show that this method provides a high recovery and good preconcentration factor. Further, the sample preparation time and consumption of toxic organic solvents are minimized without affecting the sensitivity of the method, making it novel. Thus, this method is suitable for simple and accurate determinative method for palladium in different water samples with satisfactory results as mentioned by the authors.

4.5 3,4-Dihydroxybenzaldehyde Isonicotinoyl Hydrazone (3,4-DHBINH)

Srinivas et al. [5] have developed a second-order derivative method for simultaneous determination of Pd(II) and W(VI) using 3,4-DHBINH as a complexing agent. A reaction in 1:1 molar ratio of 3,4-dihydroxybenzaldehyde and isonicotinic acid hydrazide in methanol gives 3,4-DHBINH as a product in the form of yellow crystals. Other spectrophotometric parameters have been worked out by the authors. It is described that Pd(II) reacts with 3,4-DHBINH to give a green-coloured solution at pH 3–7 and the complex of Pd(II) absorbs at 362 nm (0.53–6.40 µg/mL). A 1:1 complex is formed as determined by Job's method as well as mole ratio method. It is mentioned that Pd(II) and W(IV) interfere with each other. The method has been successfully applied for the determination of palladium in hydrogenation catalyst and tungsten in industrial waste water samples.

4.6 2,4-Dihydroxybenzophenone Oxime (DHBPO)

Lokhande et al. [6] have synthesized a new analytical reagent for the spectrophotometric determination of Pd(II), 2,4-dihydroxybenzophenone oxime (DHBPO). The reagent has been proposed as an extractant for the method. It is described that the reagent produces a yellow-coloured complex, extracted in Butan-1 at pH of 4.0. The complex gives an absorbance maxima at 410 nm and can be determined. The reported molar absorptivity and Sandell's sensitivity values are 0.672×10^2 L/mol cm and 0.06729 µg/mL. The method thus describes separation and analysis of Pd catalysed in binary mixture.

4.7 5-Methyl-2,3-hexanedione dioxime (H2MHDDO)

A new sensitive reagent for separation and extractive spectrophotometric determination of trace Pd has been developed by Tandel et al. [7]. The named reagent reacts with Pd(II) between pH 0.5 and 1.5 to give a yellow-coloured 1:2 chelate which can be easily extracted in $CHCl_3$. The concentration range following Beer's law is 0.5–9.0 µg/mL of Pd. It is described that the complex absorbs at λ_{max} 379 nm and has a molar absorptivity of

3.894×10^3 L/mol cm. It is mentioned that the method is sensitive, simple, rapid and accurate and can be successfully used for the spectrophotometric determination of Pd in synthetic alloy, ore and catalyst samples.

4.8 2,6-Dibromo-4-carboxybenzenediazo-aminoazobenzene (DBCDAA) and 1,10-phenanthroline (phen)

Guo et al. [8] have proposed a spectrophotometric method for Pd(II) determination using DBCDAA and 1,10-phenanthroline as reagents. In buffer of pH 9.7–10.6 (Na_2CO_3–$NAHCO_3$) and in the presence of p-octyl-polyethyleneglycol phenyl-ether (OP), it is explained that Pd(II) reacts with the above reagents to give a stable complex. The complex absorbs at 524 nm and the reported molar absorptivity value is 7.16×10^4 L/mol cm. The Beer's law range is up to 0.60 µg/mL of Pd and colour development takes 5 minutes after healing in 80° water bath, in which it stays for up to 24 hours. The molar ratio of Pd (II):Phen:DBCDAA is reportedly 1:1:2. It has been further stated that the method has been applied to analysis of secondary alloy with RSD of 3.8% and −5.9% error. Most of the interfering species are separable using ion-exchange chromatography as mentioned by the authors.

4.9 p-Nitroisonitrosoacetophenone

Sawant and Khambekav [9] have proposed p-nitroisonitrosoacetophenone as a new analytical reagent for the spectrophotometric determination of Pd(II). It is reported that the method uses extraction of a 1:2 (Pd:L) complex in $CHCl_3$ at pH 5.4–5.6. The determination of Pd is done at 425 nm of absorbance where Beer's law is obeyed between concentration range of 0.1 and 10 µg/mL of Pd. Reported molar absorptivity of the complex is 1.355×10^4 L/mol cm. Further, it is reported that the method is successfully applied to the determination of Pd in catalysts and ores.

4.10 1-Phenyl-1-2-butanedione Dioxime (H2PBDD)

An extraction-cum spectrophotometric determination method for Pd(II) using H2PBDD has been proposed by Tejan and Thakkar [10]. It is described that H2PBDD forms a yellow-coloured complex with Pd(II) which can be quantitatively extracted into toluene in the pH range 6.5–9.0. The extract absorbs at 380 nm in the Beer's law concentration range of 0.5–9.0 µg/mL of Pd(II). The metal to ligand ratio as mentioned is 1:2, and the molar absorptivity is 3.74×10^3 L/mol cm. It is recommended as a rapid, selective, accurate and simple method with application in synthetic and real samples.

4.11 2-Cis-3,7-di-Me 2,6-octadien-1-oxime (CDOO)

Chand et al. [11] have reported a very good method for spectrophotometric determination of Pd(II) using CDOO. It is reported that a red-coloured Pd-CDOO complex is formed at 47°–57° in 0.03–0.14 M AcOH medium. The complex is extracted into chloroform and it absorbs at 445 nm. In the concentration range of 0.4–40.0 mg/mL, the system follows Beer's law and the reported molar absorptivity is 2.434×10^3 L/mol cm with Sandell's sensitivity of 0.0007 µg/cm². The standard deviation is ±0.0012 absorbance units, and a large number of elements do not interfere. Further, the method has good reproducibility and is applicable for the analysis of industrial samples.

4.12 2,2'-Dipyridyl-2-pyridylhydrazone (DPPH)

Anthemidis et al. [12] have reported a stopped flow injection liquid–liquid extraction (SF-EX-FIA) – spectrophotometric determination method for Pd(II) determination using DPPH as an analytical reagent. It is mentioned that the complex so formed between Pd-DPPH can be extracted in $CHCl_3$ and the absorbance can be measured at 560 nm. For combining stopped flow technique with liquid–liquid extraction FI system, it is mentioned that an injection valve was used as commutator. It is described that calibration graph is linear up to 12 mg/L (Sr=0.27°/r=0.9999) with a detection limit of C_L=0.007 mg/L; sampling rate mentioned is 20 injections per hour. It is claimed that the method is suitable for the determination of Pd (II) in airborne particulate matter (APS) and automobile exhaust gas converter catalysts.

4.13 8-Aminoquinoline-5-azo-p-benzoic Acid

Zen et al. [13] have reported a spectrophotometric determination method for Pd(II) with a new analytical chromogenic reagent. It is described that the mentioned reagent in the presence of cetyltrimethyl ammonium bromide forms a blue-green-coloured complex, which absorbs at an absorbance maxima at 610 nm. The Beer's law concentration range is reported as 3.3–22 µg in 10 mL chloroform and molar absorptivity as 7.2×10^4 L/mol cm. It is reported to be applicable for the determination of palladium in ore concentration as well as catalysts successfully.

4.14   5,6-Dimethyl-1,3-indanedione-2-oxime

Rao et al. [14] have developed a new rapid, sensitive and selective method for the determination of Pd(II) using 5,6-dimethyl-1,3-indanedione-2-oxime in acetate buffer (pH 5.5). It is mentioned that the Beer's law range is 0.15–4.17 mm/mL palladium (II) and complex absorbs at 370 nm. The molar absorptivity reported is 2.98×10^4 L/mol cm. The method is successfully applied to the determination of Pd in synthetic mixtures corresponding to Pt-Ir and Oakay alloys. An abnormal shape of Job and mole ratio plots is an interesting feature of this system as described by the authors.

References

1. Shaikh, A.B., Barache, U.B., Lokhande, T.N, Kamble, G.S., Anuse, M.A. and Gaikwad, S.H. Expeditious extraction and spectrophotometric determination of palladium (II) from catalysts and alloy samples using new chromogenic reagents. *Rasayan, J. Chem.*, 2017, 10(3), 967–980.
2. Hassan, N. and Amin, A.S. Solid phase extraction and spectrophotometric determination of palladium (II) with 1-(2-benzothoazolylazo)-2-hydroxy-3-naphthoic acid. *Anal. Chem. Lett.*, 2017, 7(5), 724–736.
3. Patel, N.B. Synthesis of an analytical reagent, its spectrophotometric characterization and studies of its complexation behaviour with Pd (II) metal ion and its application. *Int. J. Appl. Pure Sci. Agric.*, 2017, 3(2), 80–88.
4. Kozani, R.R., Nakhaei, J.M. and Jamali, M.R. Rapid spectrophotometric determination of trace amounts of palladium in water samples after dispersive liquid-liquid microextraction. *Environ. Monit. Assess.*, 2013, 185, 6531–6537.
5. Srinivas, J., Kumar, A.B.V.K., Thangadurai, T.D., Rao, V.S., Yoo, Y.J. and Lee, Y.I. Non-extractive simultaneous spectrophotometric determination of trace quantities of palladium (II) and tungsten (VI). *Anal. Lett.*, 2011, 44, 815–823.
6. Lokhande, R.S., Patil, U. and Dapale, S.S. Use of 2,4-dihydroxybenzophenone oxime as a reagent for extraction and spectrophotometric determination of palladium (II). *Res. J. Chem. Environ.*, 2004, 8(4), 59–61.
7. Tandel, S.P., Jadhav, S.B. and Malve, S.P. Separation and extractive spectrophotometric determination of palladium (II) with 5-methyl-2,3-hexanedione dioxime. *Ind. J. Chem.*, 2001, 40A(10), 1128–1129.
8. Guo, Z., Du, B., Zheng, G. and Zhang, S. Spectrophotometric determination of palladium with 2,6-dibromo-4-carboxybenzene diazo-aminobenzene and 1,10-phenanthroline. *Guijinshu*, 1997, 18(4), 38–40.
9. Sawant, A.D. and Khambekav, A.M. Extraction and spectrophotometric determination of palladium (II) with p-nitroisonitrosoacetophenone. *J. Ind. Chem. Soc.*, 1997, 74(10), 824–825.
10. Tejan, A.B. and Thakkar, N.V. Extraction and spectrophotometric determination of Pd(II) with 1-phenyl-1,2-butane dioxime. *Ind. J. Chem.*, 1998, 37A(4), 364–366.
11. Chand, M., Lata, P. and Nagar, M. Spectrophotometric determination of palladium (II) using 2-cis, 3,7-dimethyl, 2,6-octadien-1-oxime. *J. Ind. Chem. Soc.*, 2003, 80(9), 861–862.

12. Anthemidis, A.N., Themelis, D.G. and Stratis, J.A. Stopped flow injection liquid-liquid extraction spectrophotometric determination of palladium in airborne particulate matter and automobile catalysts. *Talanta*, 2001, 54, 37–43.
13. Zen, Z., Zhao, Y., Xu, Q. and Zhou, J. Spectrophotometric determination of palladium with a new chromogenic reagent 8-aminoquinolune-5-azo-p-benzoic acid. *Yejin Fenxi*, 1994, 14(1), 12–14.
14. Rao, D.M., Reddy, K.H. and Reddy, D.V. Spectrophotometric determination of palladium with 5,6-dimethyl-1,3 indane-2-oxime. *Talanta*, 1991, 38(9), 1047–1050.

chapter five

Analytical Reagents Having (O), (N) and (S) as Donor Atoms and Miscellaneous Reagents

5.1 Pyridoxal Thiosemicarbazone

A non-extractive direct spectrophotometric determination of palladium (II) using pyridoxal thiosemicarbazone has been proposed by Renuka and Reddy [1]. It is described that the reagent reacts with palladium to form a pale yellow-coloured 1:3 (M:L) complex with Pd(II) in acidic medium at pH 2.0 (CH_3COONa and conc. HCl). The colour remains stable for 2 hours and can be measured at 420 nm. The reported molar absorptivity and Sandell's sensitivity are 1.63×10^4 L/mol cm and 0.635 µg/cm^2, respectively. The Beer's law concentration range or linearity graphs reported are 0.9–10.0 µg/mL of Pd.

It is claimed that this method is very successful and can be applied for the determination of Pd in hydrogenation catalysts. As reported, the complex structure and other analytical characteristics are determined and the method can be applied to Pd determination in different water samples.

5.2 4-[N'-(4-Imino-2-oxo-thiazolidin-5-ylidene)-hydrazino]-benzenesulfonic Acid (ITHBA)

A new reagent ITHBA has been synthesized and reported by Lozynska et al. [2] for determining Pd(II) spectrophotometrically. It is mentioned that the complex absorbs at $\lambda = 438$ nm in the concentration range 0.2–2.2 µg/mL of Pd(II). The reported molar absorptivity is 7.5×10^3 L/mol cm. The method reported has been studied for interference of various ions and claimed to be simple, rapid, accurate and selective as well as sensitive for Pd(II) determination. The accuracy of the method has been confirmed by voltammetric and AAS method. It is applicable to Pd determination in intermetallides and resistors.

DOI: 10.1201/9781003276418-7

5.3   5,6-Diphenyl-2,3-dihydro 1,2,4-Triazene-3-thione

Tehrani et al. [3] have reported a new spectrophotometric reagent for the determination of Pd(II). The method involves synthesis and application of the reagent in the determination of Pd(II) using this reagent. It is described that 5,6-diphenyl-2,3-dihydro 1,2,4-triazene-3-thione forms a yellow-orange coloured complex in methanol which absorbs at 385 nm. Beer's law is obeyed in the reported concentration between 10 and 50 µg/mL. The molar absorptivity reported by the authors is 6.67×10^3 L/mol cm. Further, the accuracy and precision of the method have been worked out on a within-day and between-day basis with a relative error of 3.13 and a standard deviation of 3.96. Moreover, this method has also been successfully applied to the synthetic samples and validated and compared with the results obtained with the AAS method; the difference between two methods is $p > 0.05$.

5.4   N-Decylpyridine-4-amine and Pyridine-2-thiol

A spectrophotometric determination method of Pd(II) with N-decylpyridine-4-amine in 1:1 ammonia and pyridine-2-thiol has been proposed by Bagal et al. [4]. The complex of Pd with N-decylpyridine-4-amine in 1:1 ammonia with pyridine-2-thiol can be quantitatively extracted with 10 mL of 1×10^{-4}M reagent in xylene from 0.025 M sodium malonate concentration in 25 mL aqueous phase. The study reports effect of metal ion, reagent and acid concentration as parameters. The reported molar absorptivity and Sandell's sensitivity values are 1.9×10^5 L/mol cm and 0.065 µg/cm^2, respectively.

5.5   Pyridoxal Thiosemicarbazone (PDT)

Renuka and Hussain [5] have developed a simple, highly selective and non-extractive method for spectrophotometric determination of Pd(II) using pyridoxal thiosemicarbazone as an analytical reagent. It is described that the named reagent reacts with palladium in acidic medium, pH 2.0 using CH_3COONa + concentrated (HCl) to give a yellow-coloured 1:2 (M:L) complex. Instantaneous reaction yields a complex absorbing at 420 nm which is stable for 2 hours. A linear calibration graph is obtained between 0.90 and 10.0 µg/mL of Pd(II). The molar absorptivity and Sandell's sensitivity values reported are 1.63×10^4 L/mol cm and 0.635 µg/cm^2, respectively.

5.6   (+) Cis-1,5-benzothiazepin-4-(5H) one, 3-(acetyloxy)-5-[2-(dimethylamino) ethyl]-2,3-dihydro-2-(4-methoxyphenyl)- monohydro chloride (DILTIAZEM)

Shrivastava [6] proposed a spectrophotometric method for determination of Pd(II) diltiazem complex which uses this system for its determination. This is a calcium channel blocker type of hypertensive drug which forms a stable 1:2 complex with Pd(II). λ_{max} reported is 400 nm and molar absorptivity $\varepsilon = 8.5 \times 10^2$ L/mol cm. The Beer's law obeyance is between 3.413×10^2 µg/m and 2.722×10^2 µg/mL.

5.7   P-[N,N-Bis (2-Chloroethyl)amino] Benzaldehyde Thiosemicarbazone (CEABT)

A new sensitive and selective reagent CEABT has been reported by Karthikeyan et al. [7] for spectrophotometric determination of Pd(II). A yellow-coloured complex is formed by the reagent in the pH range of 1–2 which absorbs at $\lambda_{max} = 395$ nm. Ringbom's plot method has been used to find optimum concentration range which is reported as 0.48–2.40 µg/cm^3. Molar absorptivity value as found is 4.05×10^4 dm^3/mol cm. The Beer's law range reported is ≤ 2.64 µg/cm^3, and Sandell's sensitivity of the complex from Beer's data D=0.001 is 0.0026 µg/cm^2. As reported the composition of the complex is 1:2 (M:L). It is claimed that the method can be successfully applied to determine Pd(II) in samples of alloys, complexes, catalysts, water samples as well as synthetic mixtures with accuracy.

5.8   Isonitro p-Nitroacetophenone Thiosemicarbazone (HINATS)

Barhate et al. [8] have developed an extraction-based Pd(II) determination method using HINATS. It is reported that HINATS extracts 99.6% Pd(II) quantitatively from an aqueous solution of pH 0.0–4.0 and 0.1–1 M solution of acetic or mineral acid. Further the chloroform extract has maximum absorbance at 410 nm, following Beer's law in the concentration range of 5.0–80 µg/mL of Pd(II). Reported molar absorptivity is found to be 910 L/mol cm. It is reported that interference of various ions has also been studied and the proposed method can be used for the determination of Pd(II) in catalysts.

5.9   4-(N,N-Diethylamino) Benzaldehyde Thiosemicarbazone (DEABT)

A sensitive and selective reagent for spectrophotometric analysis of Pd(II) has been proposed by Parmeshwara et al. [9]. It is reported that 4-(*N,N*-diethylamino) benzaldehyde thiosemicarbazone (DEABT) forms yellow-coloured complex with Pd(II) in KH phthalate–HCl buffer of pH 3.0. Till 3.60 µg/mL the system follows Beer's law and the optimum concentration range for minimum error as worked out from Ringbom plot method lies between 0.35 and 3.24 µg/mL. The complex shows λ_{max} at 408 nm. The reported molar absorptivity is 3.33×10^4 dm^3/mol cm, whereas Sandell's sensitivity value for D=0.001 is 0.0032 µg/cm^2. A 1:2 (M:L) complex is formed as mentioned. It is further mentioned that the study includes interferences by various ions and the method can be successfully applied to determine Pd(II) in alloys, catalysts, complexes and model mixtures accurately.

5.10   2,5-Dimercapto-1,3,4 Thiadiazole (DMTD)

A simple, selective and highly sensitive spectrophotometric method for Pd(II) determination has been proposed by Prasad [10]. In the presence of Triton X-100 in the concentration range wherein Beer's law is followed [0.20–2.5 µg/mL of Pd(II)], the complex formed shows λ_{max} at 375 nm. The reported molar absorptivity and Sandell's sensitivity values are 3.03×10^4 L/mol cm and 3.5 ng/cm^2, respectively. As reported a 1:1 complex is formed and the method is suitable for trace-level determination of palladium in alloys and minerals.

5.11   Allylthiourea

Mkrtchyan [11] reported interaction between allylthiourea and Pd(II) for spectrophotometric determination. The reported optimum determination concentration range in H$_2$SO$_4$ and HCl is within 5.0×10^{-6} to 1.4×10^{-4}M Pd(II) (2 M H$_2$SO$_4$) and 1.0×10^{-5} to 1.2×10^{-4}M Pd(II) (2 M HCl). The reported method can be applied for Pd analysis in Pd-plating electrolytes.

5.12   N,N-Diphenyl and N,N'-Diphenylthioureas

Mkrtchyan et al. [12] have studied and reported Pd(II) determination using *N,N*-diphenyl and *N,N'*-diphenylthioureas for spectrophotometric determination. It is reported that on the basis of electronic spectra, the reacting components and reaction products show formation of individual compounds. Reported concentration ranges for applicability of the method are HCl (pH 1.0–7.0 mol/L) and H_2SO_4 (pH 1.0–5.0 mol/L). The optimum conditions for the determination of Pd(II) where graduation plots are linear is 3×10^{-5} to 3×10^{-6} mol/L of Pd(II) content. The reported molar absorptivity value is $\varepsilon_{330} = 1.6 \times 10^4$ L/mol cm. A method of Pd(II) determination using above reagents has been satisfactorily applied to analysis of Pd-plating electrolyte.

5.13 1,3-Bis-(4-amino-3-mercapto-1,2, 4-triazol-5-yl) Mercaptoethane (AMTME)

An extractive spectrophotometric determination method for Pd(II) in synthetic mixtures using newly synthesized AMTME as analytical reagent has been proposed by Maldikar and Thakkar [13]. It is reported that an instantaneous formation of yellow stable complex in 1:1 ratio is observed at normal temperature in the pH range of 4.5–7.0. The complex can be selectively extracted into Me iso Bu Ketone. The observed λ_{max} is 380 nm in the Beer's law concentration range of 1.0–10.0 µg/cm^3. The reported molar absorptivity and Sandell's sensitivity values are 4.8×10^3 L/mol cm^3 and 0.0222 µg/cm^2, respectively. It is also mentioned that the study includes effect of diverse ions, pH, solvent, reagent concentration and time of equilibration. Further, as mentioned, the present method can be applied successfully for Pd determination in synthetic mixtures as well as catalyst samples.

5.14 4-Hydroxybenzaldehyde Thiosemicarbazone (HBTS)

Satheesh et al. [14] have developed a new spectrophotometric method for the determination of Pd(II) using 4-hydroxybenzaldehyde thiosemicarbazone (HBTS) as analytical reagent. It is reported that the reagent 4-HBTS gives an intense yellow colour with Pd(II) solution in acidic medium. The λ_{max} for this complex is 359 nm in a solution of pH 3.0. The reported molar absorptivity and Sandell's sensitivity values are 2.68×10^4 L/mol cm and 0.00396 µg/cm^2, respectively. The reported stability constant value of 1:1 Pd(II):4-HBTS complex is 8.98×10^5. It is also mentioned that the study includes interference studies for diverse ions.

5.15 Benzildithiosemicarbazone

Reddy et al. [15] have developed a simple yet highly sensitive method using extractive spectrophotometric technique and benzildithiosemicarbazone as reagent for Pd(II). It is reported that Pd(II) forms a reddish brown-coloured complex with the reagent in a KCl–HCl buffer at pH 2.5. The complex is easily extracted into Me iso Bu Ketone. The λ_{max} of the complex is 395 and Beer's law range of 0.25–3.5 ppm is reported by the authors. Further, the values of molar absorptivity and Sandell's sensitivity are 3.018×10^4 dm^3/mol cm and 0.0035 µg/cm^2.

5.16   4-(2-Thiazolylazo)-5-diethylaminobenzene (TDEAB)

Wang and Liu [16] have proposed a spectrophotometric determination method for Pd using TDEAB. In a mole ratio of 1:1 (Pd:R) in pH 4.0 buffer and in the presence of tetradecyl pyridinium bromide a blue complex is formed with the reagent which has a λ_{max} at 590 nm. The reported value of molar absorptivity is 1.08×105 in the Beer's law concentration range of 0–25.0 µg/25 mL. The method has been applied for the determination of Pd in Pd-C catalysts with recovery of 99% for which the results are reportedly consistent with the recommended values.

5.17   Nicotinaldehyde-4-phenyl-3-thiosemicarbazone (NPS)

A sensitive extraction-based spectrophotometric method for Pd determination has been described by Lee et al. [17] using NPS as an analytical reagent. The method incorporates formation of an insoluble palladium (II)–NPS complex in aqueous solution of pH 3.0 which can be extracted into chloroform by simply shaking for 5 minutes. The λ_{max} of the extracted complex is 365 nm and Beer's law range lies within 0.5–8.0 µg/mL of Pd(II). The reported molar absorptivity is 2.81×10^4 L/mol cm. The method is simple, sensitive and applicable to real samples of palladium.

5.18   2-Mercapto-4-methyl-5-phenylazopyrimidine

Kumar [18] has reported a method for the spectrophotometric determination of palladium, tellurium and iridium from nitric acid media, after extraction using their 2-mercapto-4-methyl-5-phenylazopyrimidine complexes into molten naphthalene. It is reported that dioxane, methanol, ethanol, 1-propanol acetone and acetonitrile were used as organic component of the mixed (polar) phase, but the maximum enhancement was obtained with acetonitrile. Solid naphthalene having complex is separated by filtering and subsequently dissolving in chloroform. This extract is used for the spectrophotometric determination. The Beer's law range reported is 7.0–84.0 µg/mL of the solution. The reported molar absorptivity and Sandell's sensitivity values are 2.05×10^4 L/mol cm and 0.0052 µg/cm^2, respectively. The mean absorbance for Pd(II) for ten replicates having 12.5 µg is 0.241 with standard deviation of 0.53%. It is reported that interference of various diverse ions has also been studied and conditions for determination of Pd in certain alloys and synthetic mixtures are also developed.

5.19 *N,N'-Dipyridylthiourea (DPyT)*

Application of *N,N'*-dipyridylthiourea (DPyT) is reported by Chaudhary and Shome [19] for the spectrophotometric determination of Pd(II) and Rh(III). It is reported that Pd(II) forms a stable complex with DPyT which shows $\lambda_{max} = 330$ in the concentration range of 0.24–2.88 µg/mL for Pd (Beer's law range). The reported molar absorptivity and Sandell's sensitivity values are 2.36×10^4 L/mol cm and 0.0065 µg/mL, respectively. It is mentioned that even in the presence of moderate excess of interfering ions the method successfully determines Pd.

5.20 *1-Phenyl-3-(3-ethyl-5-mercapto-1,2,4-triazol-4-yl)thiourea*

A simple, rapid and sensitive spectrophotometric method for the determination of Pd(II) has been reported by Shetty et al. [20]. It is reported that the mentioned reagent forms a 1:2 (M:L) complex with Pd(II) in a soluble form in aqueous medium containing 0.08–0.16 M NaOH, which shows λ_{max} at 335 nm. Up to 10 µg/mL concentration of Pd(II), the Beer's law is obeyed and the range of concentration determined from Ringbom's plot is 0.88–3.75 µg/mL.

As mentioned further, the molar absorptivity is found to be 2.42×10^4 L/mol cm and Sandell's sensitivity is 4.397×10^3 µg/cm^2. The method has been applied to the determination of Pd(II) in complexes as well as synthetic mixtures.

5.21 *Isonitrosopropionylthiophene*

A simple, rapid and selective method using isonitrosopropionylthiophene as a reagent is described by Chakkar and Kakkar [21]. It is reported that in dilute acetic acid solution, the named reagent forms a coloured complex with Pd which can be extracted into chloroform quantitatively. The reported λ_{max} for this complex is 434 nm in the Beer's law range of 0.0–6.8 µg/cm^2. The reagent reportedly forms a 1:2 complex and Sandell's sensitivity is 4.3×10^{-3} µg/cm^2. The results are reproducible and reported standard deviation is 0.58% for ten replicates using 50 µg of Pd.

5.22   Pyridopyridazine Dithione (PDP)

EI-Sayed and Abu [22] have synthesized and reported a new sensitive and selective reagent PDP for Pd(II) determination. It is reported that in a nuclear aqueous medium the named reagent forms a 1:4 complex with Pd(II). At absorbance of 570 and 615 nm, the reported molar absorptivities of the complex are 4.68×10^4 and 5.74×10^4 L/mol cm, respectively. The reported Beer's law concentration range is ≤ 2.2 and ≤ 2.5 µg/mL, with detection limits of 0.2 and 0.1 µg/mL. RSD values reported for 1.23 µg/mL are 2.6% and 1.3% in the presence and absence of micellar agent Triton X-100. It is also reported that the fourth derivative mode, the regression equation, linear range detection limit and RSD for 0.07 µg/mL are D4 = 4.3 C + 1.5×10^{-3}, 0.013–0.23 µg/m, 3.7 ng/mL and 0.78%, respectively. The reported method is free from interference of diverse ions and ionization constants of the reagent and complex stability constants have also been worked out. The successful application of the method for Pd determination in activated charcoal is an added advantage as reported.

5.23   Pyridoxal-4-phenyl-
3-thiosemicarbazone (PPT)

Sarma et al. [23] have proposed a new spectrophotometric method for Pd(II) determination using pyridoxal-4-phenyl-3-thiosemicarbazone as a reagent. It is reported as a rapid and selective extractive method for synthetic as well as samples of hydrogenation catalysts. The reagent forms a red-coloured complex at pH 3.0 with metal which is extracted into benzene. The complex absorbs at 460 nm in the Beer's law concentration range of 0.4–6.4 µg/cm^3. The molar absorptivity and Sandell's sensitivity are 2.20×10^4 dm^3/mol cm and 4.85×10^{-3} µg/cm^2, respectively. The reported correlation coefficient of the complex is 0.99, showing good linearity between the two variables. Reported detection limit is 0.05 µg/cm^3. Further, the instability constant of Pd(II)–PPT complex calculated is 2.90×10^{-5} from Edmond and Birnbaum's method and 2.80×10^{-5} at room temperature using Asmus' method. RSD for n = 5 is 1.84%. Further, it is mentioned that the method is useful for Pd(II) determination in hydrogenation catalysts as well as synthetic mixtures. The method has been further validated by AAS.

5.24   3-Thiophenaldehyde-4-phenyl-
3-thiosemicarbazone (TAPS)

A highly selective spectrophotometric method for the determination of palladium   using   3-thiophenaldehyde-4-phenyl-3-thiosemicarbazone has been proposed by Uesugi et al. [24]. An insoluble complex formed by Pd(II)–TAPs is extracted into chloroform using 1 M sulphuric acid medium and shaking for 10 minutes. The complex obeys Beer's law in the concentration range of 0.4–6.0 µg/mL of Pd and has been used for Pd(II) determination in brazing filler metals.

5.25   2-Mercaptonicotinic Acid (MENA)

Abu-Bakr [25] proposed the solution-equilibrium between Pd(II) and 2-mercaptonicotinic acid can be studied spectrophotometrically in 40% (vol/vol) ethanol–water mixture at $25 + 1°C$ and $I = 0.1 M$ $NaClO_4$. In the pH range 0.5–4.5 chelates of MENA and Pd(II) are determined by analyses of absorbance pH curves using graphical method. At 410 nm and pH 3.5 a 1:1 Pd:MENA complex has $\varepsilon = 1 \times 10^4$ L/mol cm. The effect of interfering ions has been studied, and this method is applied to the determination of Pd in catalysts as well as synthetic samples.

5.26   5-Bromosalicylaldehyde-4-phenyl-3-thiosemicarbazone (BSPS)

Uesugi et al. [26] have proposed a method based on extraction spectro-photometric determination of Pd(II) with BSPS. An insoluble complex of Pd(II)–BSPS can be extracted in chloroform from 0.05 M sulphuric acid medium. $\lambda_{max} = 412$ nm is the absorbance for measurement and the reported molar absorptivity of this complex is 1.42×10^4 L/mol cm. Up to 8.0 µg/mL of Pd(II) the complex follows Beer's law. The method has been successfully applied to the analysis of palladium using brazing filler metals.

5.27   2-N-(2-Mercaptophenyl)-1,2, 3-benzothiadiazoline (MBDT)

Watanabe [27] has used a sulphur-containing bidentate ligand MBDT synthesized by Sandmeyer in azo coupling for Pd(II) determination. The mentioned reagent selectively reacts with Pd to form an extractable complex, which is different from complexes synthesized by direct addition of reagent. The complex spectrum from that of the free reagent is red shifted. Anions such as Br^-, I^- or SCN^- effect extractability of Pd-MBTD complex, but Cl^- enhances the extraction. The concentration range of Pd determination is 0.5–40 mg Pd(II) in 10 mL of organic phase. The measurement of absorbance was done at 788 nm with a molar absorptivity of 2.9×10^4 dm^3/mol cm. The Sandell's sensitivity value reported is 3.6×10^{-3} mg Pd/cm^2.

5.28   p-Methylphenyl Thiourea (PMPT)

Kuchekar et al. [28] have described a method for spectrophotometric determination of Pd(II) using PMPT. The method is based on solvent extraction using PMPT as an extractant. A yellow complex is formed which absorbs at 300 nm, and the Beer's law range is up to 90.0 μg/mL of palladium. The reported molar absorptivity and Sandell's sensitivity values are 0.843×10^3 L/mol cm and 0.125 μg/mol cm^2, respectively. The method can be successfully applied to the determination of Pd in synthetic mixtures and alloys.

References

1. Renuka, M. and Reddy, C.O. Non-extractive spectrophotometric determination of palladium in water samples using pyridoxal thiosemicarbazone (PTSC). *Indian J. Res.*, 2017, 6(5).

2. Lozynska, L., Tymoshuk, O. and Chaban, T. Spectrophotometric studies of 4-[N'-(4-imino-2-oxo-thiazolidin-5-ylidene)-hydrazino]-benzenesulfonic acid as a reagent for the determination of palladium. *Acta Chim. Slov.*, 2015, 62, 159–167.

3. Tehrani, M.B., Tahmasebi, M., Souri, E. and Shamsha, H. Synthesis of 5,6-dipheny-2,3-dihydro-1,2,4-triazine-3-thione as a new reagent for spectrophotometric determination of Palladium. *Chem. Sci. Trans.*, 2015, 4(1), 227–233.

4. Bagal, M.R., Shaikh Uzma, P.K., Sakhre, M.A., Vaidya, S.R., Lande, M.K. and Arbad, B.R. Spectrophotometric determination of palladium (II) with N-decylpyridine-4-amine from malonate media. *J. Ind. Chem. Soc.*, 2014, 91(5), 845–851.

5. Renuka, M. and Hussain, R. Non-extractive determination of palladium in alloy samples using pyridoxal thiosemicarbazone. *Res. J. Pharm. Biol. Chem. Sci.*, 2014, 5(4), 118–130.

6. Shrivastava, R. Spectrophotometric determination of diltiazem in pharmaceutical and in-vivo samples with Pd(II). *Int. J. Pharm. Sci. Res.*, 2013, 4(12), 4676–4681.

7. Karthikeyan, J., Parmeshwara, P. and Shetty, A.N. Analytical properties of p-[N, N bis(2-chloroethyl)amino] benzaldehyde thiosemicarbazone: spectrophotometric determination of palladium (II) in alloys, catalysts and complexes. *Environ. Monit. Assess.*, 2011, 173(1–4), 569–577.

8. Barhate, V.D., Madan, P., Gupta, A.K.S. and Mandhare, D.B. Extractive spectrophotometric determination of palladium (II) with isonitroso p-nitro acetophenonethiosemicarbazone. *Orient. J. Chem.*, 2009, 25(3), 731–733.

9. Parmeshwara, P., Karthikeyan, J., Shetty, A.N. and Shetty, P. 4-(N, N-diethylamino) benzaldehyde thiosemicarbazone in the spectrophotometric determination of palladium. *Annali di Chimica (Rome, Italy)*, 2007, 97(10), 1097–1106.

10. Prasad, K.S. First derivative spectrophotometric determination of palladium (II) using 2,5-dimercapto-1,3,4-thiadiazole in neutral micelle medium. *J. Anal. Chem.*, 2006, 23(2), 27–30.

11. Mkrtchyan, A.R. Spectrophotometric determination of palladium (II) by allylthiourea. *Hayastani Kimiakan Handes*, 2006, 59(1), 58–63.
12. Mkrtchyan, A.R. Darbinyan, G.G., Shaposhnikova, G.N. and Kanchatryan, A.G. Spectrophotometric determination of palladium (II) by N, N-diphenyl- and N, N'-diphenylthioureas. *Hayastani Kimiakan Handes*, 2005, 58(4), 26–31.
13. Maldikar, A.K. and Thakkar, N.V. Extractive spectrophotometric determination of palladium (II) in synthetic and real samples using 1,3-[bis(4-amino-3-mercapto-1,2,4-triazol-5-y)mercaptooethane. *Chem. Environ. Res.*, 2004, 13(3&4), 203–211.
14. Satheesh, K.P., Rao, V.S. and Chandra Sekhar, K.B., Spectrophotometric determination of palladium using 4-hydroxybenzaldehyde thiosemicarbazone (4-HBTS). *J. Ind. Council Chem.*, 2004, 21(5), 51–54.
15. Reddy, B.K., Reddy, K.J., Kumar, J.R., Kumar, A.K. and Reddy, A.V. Highly sensitive extractive spectrophotometric determination of palladium (II) in synthetic mixtures and hydrogenation catalysts using benzildithiosemicarbazone. *Anal. Sci.*, 2004, 20(6), 925–930.
16. Wang, Y. and Liu, B. Study on spectrophotometric determination of palladium with new reagent 4-(2-thiazolylazo)-5-diethylamino benzene. *Yejin Fenxi*, 2001, 21(5), 18–20.
17. Lee, J., Uesugi, K and Choi, W. Sensitive spectrophotometric determination of palladium (II) with nicrotinaldehyde-4-phenyl-3-thiosemicarbazone. *Anal. Proceed.*, 1995, 32(7), 279–281.
18. Kumar, A. Spectrophotometric determination of palladium, tellurium and iridium after extraction with 2-mercapto-4-methyl-5-phenylazopyrimidine. *Anal. Sci.*, 1995, 11(2), 281–284.
19. Chaudhary, S.P. and Shome, S.C. Spectrophotometric determination of palladium (II) and rhodium (III) using N, N'-dipyridylthiourea as a new sensitive and selective complexing agent. *J. Ind. Chem. Soc.*, 1997, 74(7), 554–555.
20. Shetty, P., Shetty, A.N. and Gadag, R.V. Spectrophotometric determination of palladium (II) with 1-phenyl-3-(3-ethyl-5-mercapto-1,2,4-triazol-4-yl) thiourea. *Chim. Acta Turc.*, 1996, 24(2), 111–117.
21. Chakkar, A.K. and Kakkar, L.R. Extractive spectrophotometric determination of palladium using 2-(2-hydroxyimino-1-oxopropyl)thiophene. *Chem. Anal. (Warsaw)*, 1996, 1(4), 559–567.
22. El-Sayed, A.Y. and Abu, S.F.A. Spectrophotometric and derivative spectrophotometric determination of palladium (II) using pyridopyridazine dithione in the presence of non-ionic surfactant. *Mikrochim. Acta*, 1998, 29(3–4), 225–231.
23. Sarma, L.S., Kumar, J.R., Reddy, K.J., Kumar, A.K. and Reddy, A.V. A rapid and sensitive extractive spectrophotometric determination of palladium (II) in synthetic mixtures and hydrogenation catalysts using pyridoxal-4-phenyl-3-thiosemicarbazone. *Anal. Sci.*, 2002, 18(11), 1257–1261.
24. Uesugi, K., Sik, L.J., Nishoka, H., Kumagi, T. and Nagahiro, T. Extraction spectrophotometric determination of palladium with 3-thiophenaldehyde-4-phenyl-3-thiosemicarbazone. *Microchem. J.*, 1994, 20(1), 11–13.
25. Abu-Bakr, M.S. Solution equilibria of palladium (II) with 2-mercapto nicotinic acid and spectrophotometric determination of palladium. *Bull. Fac. Sci. Assiut Univ. B: Chem.*, 1993, 22(2), 123–127.

26. Uesugi, K., Nagahiro, T., Kumagi, T. and Nishioka, H. Extraction spectrophotometric determination of palladium (II) with 5-bromosalicylaldehyde 4-phenyl-3-thiosemicarbazone. *Chim. Acta Turc.*, 1991, 19(3), 245–250.
27. Watanabe, K.H.M., Imami, S. and Kobayashi, S. 2-N-(2-mercaptophenyl)-1,2,3-benzothiadiazoline as a specific reagent for the extraction – spectrophotometric determination of palladium. *Anal. Sci.*, 1989, 5, 419–423.
28. Kuchekar, S.R., Bhumkar, S. and Zaware, B. Solvent extraction and spectrophotometric determination of palladium (II) using p-methylphenyl thiourea as a complexing agent. *Int. J. Chem. Mol. Eng.*, 2017, 11(10).

chapter six

Miscellaneous Reagents for Spectrophotometric Determination of Pd(II) Having N/O/S or Other Donor Atoms

The following sections present other reagents which have not been included in chapters classified under the donor atoms yet have been used for Pd(II) determination. The reagents have N/O/S or other donor atoms as ligands.

6.1 Methylene Blue B (MBB)

Rancic et al. [1] have proposed a simple, rapid, sensitive and selective kinetic spectrophotometric method for the determination of Pd(II) in trace quantities using catalytic effect of Pd ions on the oxidation of methylene blue B (MBB) by $(NH_4)_2S_2O_8$ (APS) in citric acid buffer. It is reported that in a working wavelength of 662.4 nm and at 25°C, it is possible to determine Pd(II) traces in the concentration range of 3.3×10^{-8} to 1.0×10^{-6} g/cm^3. The RSD value was found as 2.6%–4.9% and the limit of detection was 2.0 ng/cm^3. The method has been applied to Pd(II) trace determination in alloys as well as powder of Pt. Further, the results have been compared with those obtained using ICP-OES methods with good agreement.

6.2 Cefixime

A validated spectrophotometric method for Pd(II) determination is proposed by Azmi et al. [2]. The method is based on complex formation of Pd(II) with cefixime in methanol-distilled water mixture medium in the presence of NaHPO-citric acid buffer solution of pH 6.0 at room temperature. It is reported that Beer's law is followed in the concentration range of 0.7502–16.500 µg mL. The reported molar absorptivity and Sandell's sensitivity values of the system are 1.224×10 L/mol cm and 0.002 µg/cm/0.001 absorbance unit. The limit of detection reported is 0.07, and the limit of quantification (LOQ) is 0.21 µg mL as described. Interference of ions such

DOI: 10.1201/9781003276418-8

as Cu(II), Mg(II), Mn(II), Ca(II), Fe(II), Cr(III), Ne(II), Al(III), Fe(III), Cd(II) and Zn(II) has also been studied, giving tolerance of each ion in the determination. The report mentions that the method is successfully applied for Pd(II) determination in synthetic mixtures and automobile samples from workshop areas.

6.3   Naphthol Green B

Based on catalytic effect of palladium on the oxidation of naphthol green B by periodate a simple, selective and rapid flow injection analysis (FIA) method has been proposed by Keyvanfard et al. [3] for the ultra-trace determination of palladium. It is detailed as an oxidation reaction of naphthol green B with periodate. In acidic medium, naphthol green B forms a colourless product at a very slow rate. If trace amounts of Pd(II) are present, this reaction can be done at a faster pace. The method describes that the reaction is monitored spectrophotometrically by the variation between absorbance of naphthol green B of solutions in the absence and presence of Pd(II) at the λ_{max} of 721 nm. Using these optimum conditions as established and observed it is mentioned that the influence of some essential species on the determination of palladium of flow system does not have any interfering effect on its flow injection determination. Under these optimized conditions, the absorbance signal is linearly dependent on Pd concentration in the range of 2.0–90 ng/mL with a detection limit of 0.9 ng/mL (S/N=3) and a rate of 35±5 samples/hour.

6.4   2-(5-Bromo-2-oxoindolin-3-ylidene) Hydrazine Carbothioamide, [5-Bromoisatin Thiosemicarbazone] (HBITSC)

A method for determination of Pd(II) in catalyst samples has been reported by Madan et al. [4]. It is mentioned that 2-(5-bromo-2-oxoindolin-3-ylidene) hydrazine carbothioamide, [5-bromoisatin thiosemicarbazone] (HBITSC) extracts palladium (II) quantitatively (99.90%) into n-amyl alcohol from an aqueous solution of pH 0.0–4.0 and 0.1–3.0 M HCl. An intense peak at 520 nm (λ_{max}) is observed for this extract. Further, Beer's law is obeyed for concentration range of 1.0–3.0 µg/mL of Pd(II). The reported values of molar absorptivity and Sandell's sensitivity are 7450 L/mol cm (at 520 nm) and 14.3 ng/cm². A 1:2 (Pd:HBITSC) complex composition is reported. It is further mentioned that in the proposed method, interference of various ions has also been studied.

6.5   o-Methoxyphenyl-thiourea (OMePT)

Using o-methoxyphenyl-thiourea (OMePT), a simple and sensitive method for detection of Pd(II) has been developed by Kuchekar et al. [5]. It is mentioned that trace concentration of Pd(II) can be quantitatively extracted by this reagent. A yellow-coloured Pd(II)–OMePT complex thus extracted is measured at 325 nm, which is stable for 72 hours. A 1:1 complex composition is reported by mole ratio and Job's continuous variation methods confirmed by a log–log plot. It is reported that the system follows Beer's law up to 15.0 µg/mL and molar absorptivity and Sandell's sensitivity values are 3.38×10^3 L/mol cm and 0.032 µg/cm^2, respectively. It is advantageous that the method is free from interference of cations and anions and can be applied for separation of palladium (II) from multicomponent mixtures of synthetic or alloy types.

6.6   Sodium Salt of Hexamethylene Diaminecarbodithioate (NaHMICdt.2H$_2$O)

Reported by Dhanavata et al. [6], a spectrophotometric determination method for Pd(II) using NaHMICdt.2H$_2$O is a highly sensitive, selective, simple as well as rapid one. It is reported that the reagent is a good chelating agent for palladium (II), which reacts to give a yellow-coloured complex. It can be extracted into toluene at the pH range of 0.5–2.0, and it absorbs at 435 nm (λ_{max}). The composition as reported for the complex is 1:2, and Beer's law range is 0.2–0.8 µg/mL of Pd(II). Reported values of molar absorptivity and Sandell's sensitivity are 0.754×10^4 L/mol cm and 0.0140 µg/mol cm^2, respectively.

6.7   4-Amino-3-mercapto-6-methyl-1,2,4-triazine (4H)-5-one – a N, S and O Donor Triazine

Mathew et al. [7] have reported a simple yet sensitive method for determination of Pd(II) at ppm level applying 4-amino-3-mercapto-6-methyl-1,2,4-triazine (4H)-5-one as an analytical reagent. It is reported that in aqueous media at pH 11 Pd(II) forms a light yellow-coloured metal complex with composition of ML$_2$ (Pd(II):AMMT). On the basis of Ringbom's plot, the optimum concentration range is 5.6–10 ppm of Pd(II) and the reported values of molar absorptivity and Sandell's sensitivity are 2.1121×10^4 L/mol cm and 5.3287×10^{-3} g/cm^2, respectively. The method has been studied for interference of different cations and anions as reported.

6.8   Thioglycolic Acid

A sensitive and simple spectrophotometric method for the determination of Pd(II) at ppm level has been reported by Mathew et al. [8], using thioglycolic acid as an analytical reagent. It is described that in an aqueous medium of pH 11, Pd(II) produces a light yellow-coloured complex with the mentioned reagent. The composition of the complex is ML_2 (Pd:AMMT). The reported concentration range is 5.6–10.0 ppm of Pd(II) worked out on the basis of Ringbom's plot. The molar absorptivity and Sandell's sensitivity values are 2.0692×10^4 L/mol cm and 6.8287×10^{-3} g/cm^2, respectively.

6.9   Thiomichler's Ketone (TMK)

Shemirani et al. [9] have proposed a method of Pd(II) determination by micellar extraction at the colour point temperature using TMK as a reagent (4,4'-bis(dimethylamino)thiobenzophenone). It is reported that preconcentration of 50 mL of water sample in the presence of 0.1% (w/v) octyl phenoxy polyethoxy ethanol (Triton X-114), 2×10^{-6} mol/L TMK and 1×10^{-3} mol/L buffer of pH 3 gave a detection limit of 0.47 ng/mL. The linear calibration graph is in the range of 2.50 ng/mL. Thus the method is applied to the determination of Pd(II) in natural water sample after cloud point extraction. It is mentioned that recovery of more than 97% is achieved under working condition.

6.10   Rubeanic Acid

Rahman et al. [10] have proposed a simple and selective method for trace determination of Pd(II) with rubeanic acid. It is reported that rubeanic acid in the presence of ethylenediamine (EDA), diethylenetriamine (DETA) and ammonium thiocyanate (NH$_4$SCN) shows colour in acidic medium as rubeanates. Below pH 5.0, a golden yellowish-coloured product is formed showing absorption maxima at 406 and 422 nm (ethylenediamine), 417 and 423 nm for diethylenetriamine, 415 and 425 nm (DMSO), 408 and 420 nm (NH$_4$SCN) both in direct and extracted system using iso-amyl alcohol. The molar absorptivities reported for each system are:

6724 L/mol cm for Pd(II)–RA–DETA in direct spectrophotometry
9966 mol^{-1}cm^{-1} IAA extracted Pd(II)–RA–NH$_4$SCN system

The effect of various interfering ions has also been studied and the method has been applied to Pd(II) determination in alloys and synthetic mixtures.

6.11   2,6,7-Trihydroxy-9-(3,5-dibromo-4-hydroxy) Phenylfluorone (DBHPF)

Using the mentioned reagent with cetyl trimethyl ammonium bromide (CTMAB), a new spectrophotometric method for Pd(II) determination has been reported by Guo et al. [11]. It is described that the complex formed by DBHPF in the presence of CTMAB at pH 6.1–6.7 (phosphate buffer medium) shows maximum absorption at 600 nm. The Beer's law range as reported is 0–7.0 µg Pd/10 mL, whereas molar absorptivity and Sandell's sensitivity values are 9.48×10^4 L/mol cm and 1.12×10^{-3} µg/cm², respectively. The method has been applied to synthetic sample as well as metallurgical products with prior separation of interfering species.

6.12   p-Dimethylaminobenzylidene Rhodanine in Methyl Isobutylketone

Dias et al. [12] have reported a selective and sensitive reagent for extraction spectrophotometric determination of Pd(II) in metal alloy and rock samples. The reported Beer's law range of concentration is 0.2–2.4 mg/mL of organic phase, whereas the molar absorptivity of the complex is 3.0×10^4 L/mol cm. The extraction as reported is done at pH 2.4, with a detection limit of 0.1 mg/L. A 1:1 composition of the complex is reported with no interference of various ions.

6.13   5-(5-Nitro-2-hydroxyphenylazo) Rhodanine (OP)

Ge et al. [13] have proposed a new spectrophotometric method for Pd(II) determination using 5-(5-nitro-2-hydroxyphenylazo) rhodanine (OP). In H_3PO_4 medium in the presence of OP, Pd(II) forms a stable orange red complex, which absorbs at 474 nm (λ_{max}). The reported Beer's law range is 0–45 µg/25 mL, molar absorptivity is 4.0×10^4 L/mol cm and serious interference of Au has been reported. It is mentioned that the method can be applied to the determination of Pd(II) in catalysts with RSD as 1.5%–2.1% and recovery of 99%–104%.

6.14   N-Cetyl-N'-(sodium p-aminobenzene Sulfonate) Thiourea (OPT)

Ma et al. [14] have reported palladium (II) determination method using the mentioned reagent (OPT). It is described that in the pH range of 5.0–5.6 (HAC–NaAC buffer solution), the reagent forms a yellow brown complex with 1:4 ratio, in the presence of cetyl trimethyl ammonium bromide (CTMAB). The reported λ_{max} is at 299.0 nm, and Beer's law concentration range of 2–12 μg Pd(II)/25 mL is mentioned. The molar absorptivity of $\varepsilon_{299.0}$ nm $= 1.38 \times 10^5$ L/mol cm is shown by the complex and the method is successful for Pd(II) determination in catalyst samples.

6.15   Anion Exchange Separation of Pd and Spectrophotometric Determination

Bruzzoniti et al. [15] have developed a new method for Pd(II) determination at trace levels by ion chromatographic separation and spectrophotometric detection. It is described that the technique is based on separation of Pd as $PdCl_4^{2-}$ and detection at 407 nm. The column used for separation is Pac. AS4 with HCl and $HClO_4$ as eluents. It is reported that the method is suitable for palladium determination within 300 ng/DL value.

6.16   p-[4-(3,5-Dimethylisoxazolyl) azophenylazo]calix (4) Arene

Kumar et al. [16] have proposed a new reagent p-[4-(3,5-dimethylisoxazolyl) azophenylazo]calix (4) arene for synergistic extraction and spectrophotometric determination of palladium (II) determination. It is described that from HNO_3 media after extraction the Pd, Fe and Te as complexes can be determined. For Pd, the Beer's law concentration range is 5.0–95.0 μg/10 mL and the molar absorptivity and Sandell's sensitivities reported are 1.73×10^4 L mol^{-1} cm^{-1} and 0.0061 μg cm^{-1}, respectively. The method has been successfully applied to certain alloys and synthetic mixtures.

6.17   3-Hydroxy-2-methyl-1-phenyl-4-pyridone

For microgram determination of Pd(II), a method has been developed by Vojkovic et al. [17]. The method described is based on extraction of Pd(II) by a chloroform solution of 3-hydroxy-2-methyl-1-phenyl-4-pyridone (HX). The medium which extracts a complex is aqueous H_2SO_4, in the pH range of 1.5–3.0. The extracted chloroform layer is used for Pd(II) determination. The reported molar absorptivity is 1.89×10^4 mol^{-1} dm^{3}cm^{-1}, and the absorbance is 345 nm. It is reported that no previous separation of Pd(II) and Au(III) is required if the third derivative spectrophotometry is used. The reported concentration range for determination is 0.28–8.0 μg/cm^3 even in the presence of 1.0–8.0 μg/cm^3 gold. The method has been applied to palladium determination in synthetic mixtures and Pd-charcoal.

6.18 3-Phenoxy Benzaldoxime

Lokhande et al. [18] have reported the synthesis, characterization and use of 3-phenoxy benzaldoxime for the spectrophotometric determination of Pd(II) post-extraction. It is reported that the reagent gives a yellow-coloured complex with Pd(II) which can be extracted in $CHCl_3$ at pH 4.0. λ_{max} of 435 nm is reported and Beer's law range between 0.4 and 40 µg/mL of Pd(II) is mentioned for the method. A molar absorptivity of 2.434×10^3 L/mol cm is shown by the system and the method can be applied for the Pd(II) determination in catalyst and synthetic mixtures as well. Further, the method has also been studied for interference of foreign ions.

6.19 1-(2-Pyridylazo)-2-naphtholate

Eskandari et al. [19] have reported a sensitive and selective spectrophotometric determination method for Pd(II) using modified magnetic nanoparticles and 3-phase back-extraction. It is reported that magnetite nanoparticles modified with sodium dodecyl sulphate were used to extract Pd(II) as a green 1-(2-pyridylazo)-2-naphtholate complex before its zero- and first-derivative spectrophotometric determination. The reported zero-derivative and first-derivative spectrophotometric wavelengths of absorbance are 659 and 681 nm. As reported, a sample volume of 70 mL was back-extracted with 0.50 mL of n-butanol in a 3-phase system. Reported enrichment factor is 134 and range of linearity for determination of 2–90 ng/mL is achieved and the RSDS and recovery percentages for 10 and 72 ng/mL of Pd(II) are in the range of 1.1–4.9 and 98.5%–102.6%, respectively. No significant interference of diverse ions is reported and it is further mentioned that the method has been successfully applied to Pd(II) determination in urine, alloys and palladium catalyst samples.

6.20 Optical Test Strip for Spectrophotometric Determination of Palladium Using 5-(p-Dimethylaminobenzylidene) Rhodanine

Pourreza et al. [20] have developed a new optical test strip for spectrophotometric determination of palladium (II). It is reported that the test strip is developed by mixing PVC, di-Me sebacate and 5-(p-dimethylaminobenzylidene) rhodanine in THF and coating this solution on a transparent plate using a spin device. It is reported that the effect of different variables on the strip have been studied and optimized. The calibration graph is linear in the range of 0.2–20 mg/L of palladium and the correlation coefficient is 0.9991. The detection limit is 0.1 mg/L as reported with RSD of ten replicates of 10 mg/L which is 3.1%. The method has been successfully applied to the determination of palladium in jewellery as well as synthetic samples.

6.21   Piperonal Thiosemicarbazone

Shelly et al. [21] have reported a new spectrophotometric method for Pd(II) determination using piperonal thiosemicarbazone as an analytical reagent. It is reported that this reagent forms a 1:2 (Pd:R) yellow complex soluble in 32%–40% ethanol. The complex has λ_{max} of 363 nm, and the Beer's law range is 0.5–2.45 ppm of Pd, whereas the obeyance of Beer's law is up to 3.85 ppm. Reported molar absorptivity and Sandell's sensitivity values are 3.80×10^4 L/mol cm and 2.8×10^{-3} µg/cm^2, respectively. Interference of various ions has been studied, and it is claimed that the method is very useful for palladium determination in synthetic as well as complex samples.

6.22   2-(2-Quinolinylazo)-
5-diethylaminoaniline (QADEAA)

Zhong et al. [22] have proposed a solid-phase extraction and spectrophotometric method for Pd(II) determination based on colour reaction of 2-(2-quinolinylazo)-5-diethylaminoaniline (QADEAA) and subsequent solid-phase extraction. The coloured complex is extracted with MCI-GEL reversed-phase cartridge. It is described that in the presence of 0.2–2.0 mol/L of hydrochloric acid solution and CTMAB medium, the reagent reacts with Pd(II) to yield a 1:2 stable complex. It is reported that the complex can be enriched by MCI-GEL reversed-phase cartridge and the retained complex is then eluted using acetone. Determination of Pd(II) with the eluent is done at 615 nm in the Beer's law range of 0.01–1.5 mg/L. It is successfully applied to Pd(II) determination as reported.

6.23   Novel Salen Ligand

Salhe et al. [23] have proposed a new reagent Salen, 2-[(E)-N-(2-{[2-[(E)-[(2-hydroxyphenyl)methylidene] amino]phenyl](methyl)amino}phenyl) carboxymidoyl]phenol (HHMCP), and used it for Pd(II) determination spectrophotometrically. At pH 9, the reagent forms a complex with Pd(II) which can be quantitatively extracted in chloroform. The extracted species has λ_{max}=560 nm and reported molar absorptivity is 0.7×10^2 L/mol cm. Further, it is reported to be a successful method of Pd(II) determination in synthetic mixtures and catalysts.

6.24 2-Methyl-5-(4-carboxyphenylazo)-8-hydroxyquinoline

Huang et al. [24] have reported solid-phase extraction-based spectrophotometric method using 2-methyl-5-(4-carboxyphenylazo)-8-hydroxyquinoline (MCPAHQ), a new chromogenic reagent. It is reported that a highly selective, sensitive and fast method based on solid-phase extraction of the coloured chelate on CLEAN-UPC$_{12}$ cartridge has been developed. The complex so formed has 1:2 composition which is extracted using reversed-phase C$_{12}$ cartridge. With an enrichment factor of 50, the chelate is eluted from the cartridge in minimum amount of ethanol. At 510 nm in the Beer's law range of 0.01–0.6 µg/mL, the system detects Pd(II) with a molar absorptivity value of 2.15×10^5 L/mol cm. The reported value of RSD for 11 replicates having 0.8 µg/L is 2.2%. A detection limit of 0.1 µg/L is mentioned, and the method has been applied to trace determination of Pd(II) in automobile exhaust, converter catalysts, etc.

6.25 Phthalaldehyde Acid Thiosemicarbazone (PAATSC)

A method for extraction spectrophotometric determination of Pd(II) using a new thiosemicarbazone reagent PAATSC has been developed by Salmas et al. [25]. It is reported that the mentioned reagent forms a yellow complex with Pd(II) which is extracted from an aqueous solution of pH 4.0 (HAC/NaAC 2 mol/L) medium. The complex absorbs at 355 nm, and apparent molar absorptivity is $(\varepsilon) = 5.1 \times 10^4$ L/mol cm. The detection limit is 23 ng/mL, and Beer's law range of 0.1–2.0 µg/mL of Pd(II) is mentioned.

6.26 Flow Injection Analysis and Spectrophotometric Determination of Palladium

Shrivas et al. [26] have developed a simple and specific method for flow injection analysis (FIA) spectrophotometric determination method for Pd(II) determination. It is based on colour reaction of Pd(II) with N-phenylbenzimidoyl thiourea (PBITU) in the acidic range of 0.2–2.0 M HCl in 10% (v/v) ethanolic solution. The reported molar absorptivity at 345 nm is 1.20×10^4 L/mol cm. The method can be applied for the determination of Pd in both catalytic and synthetic mixtures.

6.27   2-Aminoacetophenone
Isonicotinoylhydrazone (2-AAINH)

Gangadharappa et al. [27] have developed a sensitive and selective reagent for spectrophotometric determination of Pd(II) using 2-AAINH as reagent. An intense orange red complex is formed by the reagent in acidic pH. Complex absorbs at 500 nm and pH range is 3–5. It is reported that Beer's law is obeyed in the range of 0.30–3.00 µg/mL of Pd(II) at pH 4. The reported value of molar absorptivity is 3.00×10^4 L/mol cm and Sandell's sensitivity as reported is 0.0035 µg/cm². A 1:2 (Pd–2-AAINH) complex is formed. The first-derivative spectrum has a peak at 540 nm with a zero cross at 494 nm, whereas the second-derivative spectrum has λ_{max} of 560 nm and a minima at 500 nm, with a zero cross at 537 nm. This shows that a sensitive first- and second-derivative spectrophotometric method is also proposed for Pd(II) determination, as reported.

6.28   Crystal Violet with Hypophosphite

Jabbari et al. [28] have developed a palladium (II) determination technique based on the catalytic-kinetic method using catalytic effect of Pd(II) on reaction of crystal violet with hypophosphite. It is described that the reaction is monitored spectrophotometrically at 590 nm over the concentration range of 0.071–2.850 ppm of Pd(II). Reported detection limit is 0.071 ppm with a standard deviation of 2.76%. It is reported that the proposed method has been applied to the determination of Pd(II) in river water and Pd catalysts. Further, it has also been used for the determination of propylthiouracil in pharmaceutical samples using its inhibitory effect of Pd(II)-catalysed reaction between crystal-violet and hypophosphite ions. It is stated that the decrease in absorbance of crystal violet at 590 nm is proportional to 1.7–248.0 ppb of polythiouracil (PTU).

6.29   O-Methylphenylthiourea

Shelar et al. [29] have proposed a solvent extraction-based spectrophotometric determination method for Pd(II) using low concentration of O-methylphenylthiourea (OMPT). It is reported that the mentioned reagent in chloroform extracts trace concentration of Pd(II) at 0.8 mol/dm³ HCl media, requiring only 10 seconds equilibration time.

The extracted species is 1:1, stable for >70 hours and has absorbance at 340 nm. The Beer's law range reported is 0.01–150 µg/cm³. The values of molar absorptivity and Sandell's sensitivity are 2.85×10^3 dm³/mol/cm and 0.037 µg/cm², respectively. With no interference from a large number of cations and anions, it is applicable for separation of Pd(II) from multi-component mixtures as well as hydrogenation catalysts.

6.30 Diacetylmonoxime-(p-anisyl)-thiosemicarbazone

Varghhese et al. [30] have developed a new reagent diacetylmonoxime-(p-anisyl)-thiosemicarbazone for the determination of Pd(II). The complex absorbs at 440 nm in the Beer's law range 0.2–2.0 µg/mL of Pd. The reported molar absorptivity and Sandell's sensitivity values are 3.8×10^4 L/mol cm and 2.8 ng/cm². It is further mentioned that calibration graph for first-derivative spectrophotometric determination of Pd(II) is derived by measuring derivative amplitudes at 48 nm in the linear range of 0.15–2.6 µg/mL, and most of the metal ions associated with Pd in mineral or alloy samples do not interfere. As mentioned, the method can be successfully applied to Pd determination in hydrogenation catalysts and synthetic mixtures.

6.31 2-Hydroxy-3-methoxybenzaldehyde Thiosemicarbazone (HMBATSC)

Srivani et al. [31] have proposed a newly synthesized reagent HMBATSC for rapid, simple, selective and sensitive method for spectrophotometric determination of Pd(II). It is reported that Pd(II) forms a yellowish green-coloured M_2L_3 complex in DMF with the mentioned reagent in the pH range between 1.0 and 7.0 at room temperature. The complex so formed shows λ_{max} at 380 nm in the Beer's law range of 0.426–44.257 µg/mL of Pd. The reported molar absorptivity and Sandell's sensitivity are 2.198×10^4 L/mol cm and 0.049 µg/cm². A second-order derivative method has also been proposed by the authors. It is mentioned that the method can be used for the determination of Pd in alloy steels and hydrogenation catalysts.

6.32 Hexylbenzimidazolylsulfide (HBMS)

Huang et al. [32] have prepared and used hexylbenzimidazolylsulfide, a new chromogenic reagent for determination of Pd(II). It is reported as a highly sensitive, selective and fast method which is based on a rapid reaction of Pd(II) with HBMS and subsequent solid-phase extraction of the complex on the CLEAN-UPC8 cartridge. It is reported that in the presence of 0.01–0.1 m/L HCl solution and polyoxyethylene–nonylphenol ether (emulsifier – OP) medium HBMS forms a coloured complex with Pd in the molar ratio of 1:2. Crystal structure based on X-ray diffraction has been described. Thus, the complex formed can be eluted from the cartridge with DMF with an enrichment factor of 50. The reported molar absorptivity of the complex is 2.08×10^5 L/mol cm at 452 nm. The Beer's law range is

0.01–0.6 µg/mL. Detection limit of 0.1 µg/L is reported and RS for 11 replicates of 0.001 µg/mL is 2.8%. The proposed method has been applied to the determination of Pd traces in automobile exhaust gas converter catalysts successfully.

6.33   p-Sulfobenzylidene-Rhodanine

A new method for Pd(II) in cyanide residue is proposed by Uang et al. [33]. The method is based on colour reaction of p-sulfobenzylidene-rhodanine with Pd(II) and subsequent solid-phase extraction with C_{18} cartridge. It is reported that in pH 2.0 buffer solution medium in the presence of CTMAB, the reagent reacts with Pd making a stable 2:1 complex. The reported molar absorptivity is 7.79×10^4 L/mol cm at 535 nm. The reported Beer's law range is 0.01–2.0 µg/mL. It is further mentioned that the method has been successfully applied for the determination of Pd in cyanide residues.

6.34   3,5-Dimethoxy-4-hydroxybenzaldehyde isonicotinoylhydrazone

Rao et al. [34] have developed a sensitive derivative spectrophotometric determination method for palladium (II) using 3,5-dimethoxy-4-hydroxy-benzaldehydeisonicotinoylhydrazone in the presence of micellar medium of Triton X-100, a neutral surfactant. A bright yellow-coloured complex is formed which is stable, water soluble with a $\lambda_{max} = 382$ nm. The pH of buffer medium is reported as 5.5, in the Beer's law range of 0.1064–2.1284 µg/mL of Pd(II). The molar absorptivity, Sandell's sensitivity and stability constants as reported by the authors are 2.44×10^4 L/mol cm, 0.0044 µg/cm^2 and 2.2×10^7, respectively. The method has been used for Pd(II) determination in real water samples, hydrogenation catalysts as well as alloy samples successfully.

6.35   1,3-Bis(hydroxymethyl) benzimidazole-2-thione

Gaekwad et al. [35] have proposed a rapid and highly sensitive extraction-based method using 1,3-bis(hydroxymethyl)benzimidazole-2-thione as an analytical reagent for Pd(II) determination. It is described that BHMBT in Me isobutyl ketone at pH 0.7–3.5 M, perchloric acid medium forms a yellow complex with Pd(II) at 370 nm. Beer's law is followed up to 600 ppm of Pd(II) in the concentration range of 1.32–5.01 ppm. A 1:1 complex is formed having molar absorptivity as 1.53×10^4 L/mol cm and a Sandell's sensitivity of 0.0068 µg/cm^2. Applicability of the reported method has been mentioned in the Pd determination of synthetic mixtures of alloys and hydrogenation catalysts.

6.36   2,6-Diacetylpyridine bis-4-phenyl-3-thiosemicarbazone (2,6-DAPBPTSC)

Reddy et al. [36] have developed a new reagent for separation and determination of Pd(II). It is reported that the new reagent forms a yellowish, orange-coloured 1:1 chelate at pH 4.0 which can be easily extracted into iso amyl alcohol. The complex absorbs at 410 nm in the Beer's law range of 0.0–12.6 µg/mL with correlation coefficient value of 0.962. The parameters including molar absorptivity, Sandell's sensitivity and stability constant have been worked out and reported as 1.156×10^4 L/mol cm, 0.0092 µg/cm^2 and 1.667×10^4 (Asmus method). Standard deviation (n=5) is reported as 0.371. It is reported that determination of Pd(II) in spiked samples can be successfully done by this method.

6.37   2-Hydrazinopyridine

Soliman et al. [37] have developed a rapid, simple, sensitive and validated spectrophotometric method for determination of Pd(II) using 2-hydrazinopyridine. It is reported that the complex formed with the reagent can be measured at 510 nm. Under optimum conditions, a purple-coloured complex with a 1:1 stoichiometry in the Beer's law concentration range of 1.06–9.00 µg/mL is obtained. The reported molar absorptivity is 2.978×10^3 L/mol.cm with RSD % (0.04–0.41) and percentage recovery of 96.61–102.58. The interference of diverse ions has also been studied as mentioned.

6.38   4-(5-Chloro-2-pyridine)-azo-1,3-diaminobenzene

Qu et al. [38] have studied chromogenic reaction of 5-Cl.PADAB with Pd. It is mentioned that in 1.80 mol/L sulphuric acid medium, the reagent forms a coloured complex with Pd absorbing at 570 nm. The system followed Beer's law in the range of 0.0–1.2 mg/L of Pd concentration. It has been mentioned that no interference was observed with 30 ions, except Pb^{2+}, Ag^+ and Cr(VI), which could also be eliminated by precipitating Pb^{2+}, Ab^+ with sulphuric acid and HCl and anhydrous ethanol was added. Subsequently, Cr(VI) was reduced to Cr(IV) by heating. Reported molar absorptivity is 6.38×10^4 L/mol cm. The method has been successfully applied to common ore samples for Pd determination.

6.39   Bromosulfonazo III

Based on chromogenic reaction of Pd(II) with bromosulfonazo (III) in a medium of pH 2.9 buffer solution a method has been developed by Yuan et al. [39]. It is reported that a blue-coloured complex absorbing at 623.7 nm has molar absorptivity of 6.375×10^4 L/mol cm. A Beer's law range of 0.32 µg Pd in 25 mL solution is also reported. The detection limit of 2.132 µg/L for Pd is achieved and the proposed method has been applied to the determination of Pd-catalyst whose results are consistent with those obtained using the AAS method.

6.40   2-(3,5-Dichloro-pyrdylazo)-5-dimethylaminoaniline (3,5-diCl-PADMA)

Li et al. [40] have proposed a new solid-phase extraction-based spectrophotometric method for Pd determination with MCGEL CHP 20Y as sorbent. It is described that 3,5-diCl PADMA reagent forms a complex with Pd(II) rapidly which is extracted with reversed-phase MCI-GEL CHP 2OY resin. In the presence of 0.05–0.50 mol/L of HCl and CTMAB medium a 1:3 (Pd:L) purple chelate is formed. The complex is enriched by the solid-phase extraction with MCI-GEL CHP 2OY cartridge. The retained chelate is eluted with ethanol. Beer's law is obeyed in the concentration range of 0.01–3.0 µg/mL of the solution. The reported detection limit is 0.05 µg/L and the method has been successfully applied to Pd determination.

6.41   3-Hydroxy-3-propyl-1-(4-carbamidosulfamoyl)phenyltriazene (CSPT)

Salvi et al. [41] have developed a new spectrophotometric method for determination of Pd(II) using above reagent. It is reported that the reagent forms a light violet-coloured complex in alcoholic medium at pH between 1.8 and 2.0. The reported molar absorptivity and Sandell's sensitivity values are 8372 L/mol cm and 12.71 mg/cm for a 1:2 (M:L) complex.

6.42   Ion Floatation Spectrophotometric Determination

Dong et al. [42] have proposed an ion floatation spectrophotometric determination method for Pd(II) determination in acetaldehyde catalysts using 1-(2-pyridylazo)-2-naphthol as collector. It is described that using PAN as collector, with suitable pH control and determination of wavelength Pd can be determined without separation from Cu(II) in acetaldehyde catalyst. The relative error of the method is 0.61% and RSD is 2.33% with satisfactory results as mentioned.

6.43 2-(2-Quinolinoylazo)-1, 3-diaminobenzene (QADAB)

Lin et al. [43] have reported the synthesis and application of 2-(2-quinolinoylazo)-1,3-diaminobenzene for spectrophotometric determination of Pd(II). It is detailed as in the presence of 0.2–3.0 mol/L perchloric acid QADAB reacts with Pd(II) to give a stable 2:1 complex at 590 nm. The Beer's law range is 0.01–1.2 mg/L, and the molar absorptivity value is 8.08×10^4 L/mol cm.

6.44 p-Sulfobenzylidinerhodanine (SBDR)

Yang et al. [44] have proposed a spectrophotometric method of Pd(II) determination using p-sulfobenzylidinerhodanine as an analytical reagent. It is reported that in HCl solution SBDR, the reagent reacts with Pd to give a 2:1 stable coloured complex. The reported wavelength is 530 nm and Beer's law range is obeyed between 0.2 and 50 µg/25 mL of Pd. The molar absorptivity is 5.79×10^4 L/mol cm and the method has been successfully applied to the determination of Pd(II) in catalysts with fairly good results.

6.45 5-Iodo-[calix(4)arene] aminoquinoline (ICAQ)

Huang et al. [45] have reported a new reagent 5-iodo-[calix(4)arene]aminoquinoline (ICAQ) for Pd(II) determination. It is reported that the synthesized reagent ICAQ reacts with Pd(II) in the presence of Triton X-100 to form a stable complex in the ratio of 1:1 (L:M) at pH 10.2–11.0 (Na_2CO_3–$NAHCO_3$ buffer). λ_{max} is reported as 625 nm and the Beer's law range is 0–0.80 µg/mL for Pd(II). The reported molar absorptivity is 1.42×10^5 L/mol cm. It is further mentioned that RSD (n=6) with precision of 2.5%–3.0% is comparable to AAS results.

6.46 2,4-Dichlorophenylfluorone (DCIPF)

As reported by Chen et al. [46], the reagent DCIPF rapidly reacts with Pd(II) to form a stable complex in the ratio of 1:5. This is reported at room temperature and in weakly acidic medium. The Pd is determined at 470 nm in HAC–NaAC buffer in the presence of 1.7–2.6 mL of DCIPF and

CTMAB as surfactant. The reported molar absorptivity is 4.46×10^5 L/mol cm for Pd determination in Pd catalysts. It has been mentioned that the method can be applied to the determination of Pd in Pd-mineral and Pd catalysts with good results.

6.47 2-(5-Carboxy-1,3,4-triazolylazo)-5-diethylaminoaniline (CTZAN)

Chen et al. [47] have reported optimum conditions for Pd(II)-CTZAN colour reaction with Pd(II) for its determination spectrophotometrically. It is reported that the mentioned reagent gives a stable purple red complex (2:1) in the buffer of pH (Ac-NaAc) 5.0. The complex has 585 nm absorbance value and molar absorptivity is 1.2×10^4 L/mol cm. The Beer's law obeyance range mentioned is 0–1.6 mg/L for Pd(II).

6.48 4-(2-Pyridylazo) Resorcinol and 1-(2-Pyridylazo)-2-naphthol

Shavkunova et al. [48] have developed a determination method for Pd(II) using 4-(2-pyridylazo) resorcinol and 1-(2-pyridylazo)-2-naphthol as analytical reagents. It is reported that micro quantities of palladium are determined by spectrophotometric method using the two reagents. The solution is obtained by HCl leaching of potassium production by products subjected to chlorinating calcination. It is mentioned that this model solution was studied and the effect of co-solutes on the determination of Pd in leaching solution was also studied. This way content of Pd in samples was determined.

6.49 2-Hydroxy-3-nitroso-5-methyl Acetophenone Oxime (HNMA)

Lokhande et al. [49] have developed an extractive spectrophotometric method for Pd(II) determination using HNMA as a reagent. The HNMA extracts of Pd(II) into chloroform from an aqueous solution of pH 0.0–4.0 and from 0.1 to 1 M acetic acid and mineral acid quantitatively has 99.96% of Pd. It is reported that the extract shows an intense peak at 430 nm (e_{max}). The Beer's law range is 0.1–10 µg/cm^3 and molar absorptivity is 15796 dm^3/mol cm at 430 nm. A 1:2 (Pd:HNMA) complex is reported and interference of various ions has also been studied. The method is reportedly useful for Pd(II) in catalyst samples.

6.50   4-(2-Thiazolylazo)-5-diethylamine-aminobenzene

Wang et al. [50] have proposed a new method for the spectrophotometric determination of Pd using 4-(2-thiazolylazo)-5-diethylamine-aminobenzene as a reagent. It is reported that the reagent forms a blue complex in acetate buffer of pH 4.0 in the presence of tetradecyl pyridinium bromide with Pd(II). The complex has a $\lambda_{max}=590$ nm. The reported Beer's law range is 0–25.0 µg/25 mL and molar absorptivity is 1.08×10^5 L/mol cm. The method has been used for Pd determination in Pd catalysts.

6.51   Nanocatalytic Spectrophotometric Determination of Palladium (II) Using $NiCl_2$- NaH_2PO_4 System

Tang et al. [51] have developed a nanocatalytic spectrophotometric method for Pd determination using $NiCl_2$–NaH_2PO_4 system. It is reported that in NaOH medium at 85°C reduction of $NiCl_2$ by NaH_2PO_2 to NIP particles is slow. It is also reported that the absorbance is weak and blank solution is colourless and transparent. When palladium (II) is added as a catalyst to this reaction it accelerates the reaction which enhances the absorbance to 395 nm. The dark colour is obtained. On this basis a catalytic spectrophotometric method has been developed for Pd(II) determination. The linear range of detection is 0.09–1.80 µg/L of Pd (II) with 0.03 µg/L Pd as detection limit. Thus, it is claimed to be a simple, selective and highly sensitive method by the authors.

6.52   p-Iodochlorophosphonazo-Tween-60

Liu et al. [52] have proposed a spectrophotometric method for the determination of micro amount palladium which is based on reaction between palladium and p-iodochlorophosphonazo in the presence of Tween-60. The tertiary system as reported absorbs at 638 nm and Beer's law range is 0.10–1.20 mg/L. The reported value of molar absorptivity is 6.0×10^4 L/mol cm. It is a selective method and presence of Au^{3+} and Ag^+ up to five times is tolerated. The results of trace palladium determination in catalytic recovery as well as secondary anode slime are satisfactory as reported.

6.53　Solid-Phase Transmission Spectrophotometric Determination of Palladium

Wu et al. [53] have reported a solid-phase transmission spectrophoto-metric method for the determination of palladium using PVC functional membrane. It is reported that Pd(II) reacts with thiomichler's ketone, which is distributed homogeneously in the PVC functional membrane by forming a hydrophobic compound. The medium of the reaction is HOAc–NaOAc of pH 3–4. This hydrophobic compound can be detected using self-made solid-phase transmission attachment at λ_{max} 520 nm. The reported Beer's law range is 0–7.5 μg/10 mL with a detection limit of 0.015 μg/10 mL. Consistent results are obtained for Pd in minerals using this method.

6.54　4-Hydroxy-1-Naphthalrhodanine (HNR)

Wu et al. [54] have developed a spectrophotometric method for Pd deter-mination using colour reaction of 4-hydroxy-1-naphthalrhodanine (HNR) and solid-phase extraction of the coloured complex using C_{18} cartridge. It is reported that in pH 3–5 HOAc–NaOAc buffer medium with Tween 80, the reagent reacts with Pd to form a stable 2:1 complex which is enriched by C_{18} cartridge. The retained chelate is then eluted with EtOH. The elute has an absorbance at 520 nm in the Beer's law concentration range of 0.01–2.0 mg/L. The method is applied to determine Pd in cyanide residues successfully.

6.55　2-(2-Quinolinylazo)-5-dimethylaminobenzoic Acid (QADMAB)

Zhu et al. [55] have proposed a new method for spectrophotometric deter-mination of palladium. The method is based on the colour reaction of the reagent 2-(2-quinolinylazo)-5-dimethylaminobenzoic acid (QADMAB) with Pd(II) and extraction of coloured complex in solid phase with C_{18} car-tridge. It is reported that in the presence of pH 3.5 buffer (HOAc–NaOAc) and CTMAB, the reagent forms a 2:1 complex. The obtained coloured complex is enriched by C_{18} cartridge and subsequently eluted in EtOH. This is determined at 630 nm in the concentration range (Beer's law) at 0.01–1.5 mg/L using spectrophotometry. The method has been success-fully applied for Pd determination in cyanide residues.

6.56 2-(p-Carboxyphenylazo)benzothiazole

A method based on solid-phase extraction and spectrophotometric determination has been proposed by Huang et al. [56]. The method uses colour reaction of 2-(p-carboxyphenylazo)benzothiazole (CPABT) with palladium (II) and subsequent extraction of its coloured complex with C_8 cartridge. It is reported that in the presence of sodium dodecylbenzene sulfonate (SDBS) and KH_2PO_4–K_2HPO_4 buffer of pH 5.9–7.5 CPABT reacts with palladium (II) to form a stable 1:1 complex. This complex is extracted with C_8 cartridge and further eluted from cartridge with methanol. This can be determined at 510 nm in the Beer's law range of 0.1–1.0 µg/mL. The reported molar absorptivity is 1.61×10^5 L/mol cm. The method as reported can be successfully applied to determining trace palladium in catalysts.

6.57 2-(2-Quinolylazo)-5-Diethylaminobenzoic Acid (QADEAB)

Yang et al. [57] have developed a sensitive, selective and rapid method for determination of Pd based on the reaction of Pd(II) with the ligand 2-(2-quinolylazo)-5-diethylaminobenzoic acid. The chelate so formed is extracted with a reverse-phase polymer-based C_{18} cartridge. It is reported that in the presence of 0.05–0.5 mol/L of HCl solution and cetyl trimethyl ammonium bromide (CTMAB) medium the reagent QADEAB gives a violet complex with Pd in a molar ratio of 1:2 (Pd:QADEAB). The violet complex is enriched using C_{18} cartridge and an enrichment factor of 200 is achieved by elution of the chelate. It is then eluted in minimum quantity of isopentyl alcohol. The Pd can be determined now at 628 nm, in the Beer's law range of 0.01–1.2 µg/mL. The reported molar absorptivity is 1.43×10^3 L/mol cm. The reported RSD for 11 replicate samples is 2.18% at the 0.2 µg/L level. A detection limit of 0.02 µg/L is achieved as claimed. The method has been successfully applied to Pd determination in environmental samples.

6.58 4-(5-Chloro-2-pyridyl)-azo-1,3-diaminobenzene

Hu et al. [58] have studied colouring reaction of Pd(II) with 4-(5-chloro-2-pyridyl)-azo-1,3-diaminobenzene (5-Cl-PADAB) for spectrophotometric determination. It is described that in the presence of 3.0 mol/L H_3PO_4 at 100°C, the reagent forms a stable 2:1 complex which has λ_{max} at 570 nm. The reported molar absorptivity is 6.39×10^4 L/mol cm in the Beer's law range of 0–1.2 mg/L for Pd. Compared to the results of AAS, the determination of Pd using this method in minerals is in quite good agreement. Standard deviation is less than 4% with recoveries of samples between 98.5% and 103.0%.

6.59   N-Decylpyridine-4-amine

Bagal et al. [59] have developed a spectrophotometric determination method for palladium (II) using N-decylpyridine-4-amine from malonate medium. It is reported that Pd(II) reacts with the reagent in xylene from malonate medium to form a complex which is extracted with 10 mL of 1×10^{-4}M reagent concentration in xylene from 0.025 m sodium malonate in 25 mL aqueous phase with 1:1 ammonia. This is then estimated spectrophotometrically with pyridine-2-thiol at 410 nm. Effect of diverse ions has been studied and it is mentioned that this is a highly selective, simple and reproducible method. The reported values of molar absorptivity and Sandell's sensitivity are 1.9×105 L/mol cm and 0.065 µg/cm^2.

6.60   2-(5-Carboxy-1,3,4-triazolyl-azo)-5-diethylaminobenzoic Acid (CTZDBA)

Zhu et al. [60] have proposed a spectrophotometric method for determination of Pd(II) using CTZDBA as a reagent to form a 1:2 stable, purple complex; the absorption maxima λ_{max} is reported to be at 548 nm. Further, the molar absorptivity of the complex is 9.32×10^4 L/mol cm in the Beer's law range between 0.08 and 0.8 mg/L. The method has been applied to Pd(II) determination in Pd catalysts supported on activated charcoal and carbon nanotube samples.

6.61   1-(2'-Benzothiazole)-3-(4'-carboxybenzene)triazene (BTCBT)

Huang et al. [61] have developed a spectrophotometric method of Pd(II) determination using its colour reaction with 1-(2'benzothiazole)-3-(4'-carboxybenzene)triazene. This complex is then extracted with C$_8$ cartridge using solid-phase extraction method. It is reported that in the presence of OP-SDBS and C$_3$H$_5$(COO)$_3$HNa$_2$–NaOH buffer (pH 5.0–6.3) as medium, the reagent reacts with Pd(II) to give a stable 2:1 complex, which is extracted on C$_8$ cartridge and further eluted with ethanol. The eluted complex is used for determination at 490 nm. The reported Beer's law range is 0.1–1.2 µg/mL with molar absorptivity of 1.16×10^5 L/mol cm. As reported the method is successfully applied to trace palladium determination in C-Pd catalysts.

6.62   Tween 80-(NH$_4$)$_2$SO$_4$-PAR System

Using Tween 80-(NH$_4$)$_2$SO$_4$-PAR system, Wang et al. [62] have proposed a new spectrophotometric method for the determination of Pd(II). It is reported that on the basis of separation behaviour of Pd(II), Rh(II) and Pt(IV), the liquid–solid extraction is done using macromolecule Tween-80 water solution with water-soluble chelated agent PAR as extractant. (NH$_4$)$_2$SO$_4$ is used as phase separation salt in the medium adjusted by EDTA–NaOH solution. The extracted Pd(II) reagent species shows molar absorptivity ε_{528} as 4.2302×10^4 L/mol cm and the method has detection limit of palladium as 0.026 µg/10 mL. Further, the Beer's law range for determination is 23.2 µg/10 mL of Pd. The authors claim that the method is good in selectivity with no interference of common anions and cations and can be satisfactorily used for natural and synthetic samples.

6.63   1-(2-Pyridylazo)-2-naphthol (PAN)

Koosha et al. [63] have proposed a new method on the foundation of optimizing air-assisted liquid–liquid microextraction by the Box–Behnken design and spectrophotometric determination of palladium in water samples. It is reported that this fast, simple pre-treatment technique is based on air-assisted liquid–liquid microextraction (AALLME). Using 1-(2-pyridylazo)-2-naphthol (PAN) as a chelating agent and chloroform as extractant, various parameters are optimized in the study using Box–Behnken design. The linear concentration range is 5.700 ng/mL with correlation coefficient (r^2) as 0.996. Various parameters as reported are detection limit 0.7 ng/mL, enrichment factor of 193, and intra- and inter-day precision as 1.3% and 1.8%. Further, it is mentioned that this method can be successfully applied to trace determination of palladium in environmental water samples.

6.64   (p-Methoxyphenyl) ethane-1,2-Dione-1-Oxime

A simple yet selective method for spectrophotometric determination of palladium (II) has been proposed by Narhari et al. [64]. The method as described is based on the colour reaction of Pd(II) with (p-methoxyphenyl) ethane-1,2-dione-1-oxime (HMPEDO) which yields a yellow-coloured 1:2 complex, extracted into chloroform in the pH range 1–20 µg/cm^3. The complex can be measured at 420 nm (λ_{max}) and obeys Beer's law in the concentration range of 1–20 µg/cm^3. The instability constants of the complex have been calculated by the Asmus method which is 2.76×10^{-4} at room temperature, as well as using mole ratio method to be 2.68×10^{-4}. The method has been successfully applied for the determination of palladium in both synthetic as well as real samples.

6.65 4-(N'-4-Imino-2-oxo-thiazolidine-5-ylidene)-hydrazinobenzoic Acid (p-ITYBA)

Tymoshuk et al. [65] have developed a method for palladium (II) ions using 4-(N'-4-imino-2-oxo-thiazolidine-5-ylidene)-hydrazinobenzoic acid (p-ITYBA) as an analytical reagent. The ligand forms a complex with Pd(II) in water medium showing maximum absorbance at 450 nm. The pH of the optimum complex formation is 7.0 and the molar absorptivity is 4.30×10^3 L/mol cm. It has a Beer's law range of 0.64–10.64 µg/mL for Pd(II). Advantages of the method claimed are that it is simple, rapid and with no interference of Co(II), Ni(II), Zn(II), Fe(II), Cu(II), Al(III) and other ions. The method has been successfully applied to the determination of Pd in catalysts and other objects.

6.66 Simultaneous Injection Effective Mixing Flow Analysis System of Palladium Determination Spectrophotometrically

Quezada et al. [66] have proposed a simultaneous injection effective mixing flow analysis system (SIEMA) for the spectrophotometric determination of palladium (II). The colour reaction which is used for the method is between Pd(II) and 2-(5-bromo-2-pyridylazo)-5-[N-n-propyl-N-(3-sulfopropyl)amino]aniline (5-Br-PSAA). The reaction gives a blue complex absorption at 612 nm. This is claimed that analysis performance of the proposed method is superior to a conventional FIA method using 5-Br-PSAA in various parameters like reagent consumption, waste volume and time of analysis. The method has been applied to the determination of palladium in dental alloys and hydrogenation catalysts.

6.67 N-Amyl-N'-(sodium p-aminobenzenesulfonate)thiourea (APT)

Ma et al. [67] have developed a highly sensitive spectrophotometric reagent for palladium determination. It is reported that in the presence of cetyl trimethyl ammonium bromide (CTMAB), the named reagent APT forms a yellow stable complex in HAc–NaAc medium. The complex absorbs at 294.4 nm and has an apparent molar absorptivity of 3.52×10^5 L/mol cm.

Advantageously the reagent is water soluble and can be used for spectrophotometric determination of palladium in various samples including minerals and catalysts.

6.68   p-Dimethylaminobenzaldehyde-(p-anisyl)-thiosemicarbazone (DBATSC) and Glyoxal (p-anisyl)-thio Semicarbazone (GATSC)

Two new chromogenic reagents DBATSC and GATSC have been developed by Pasha et al. [68] for the spectrophotometric determination of Pd(II). It is mentioned that the two reagents react with palladium (II) to make a coloured complex. The method facilitates accurate determination of palladium by both direct and derivative spectrophotometry. The methods have been claimed to be simple, selective as well as sensitive in aqueous medium and do not need prior separation. Using the methods, palladium (II) can be determined in synthetic as well as hydrogenation catalyst samples.

6.69   Food Additive Dye (Yellow-6)

Leon-Rodriguez et al. [69] have proposed a simple spectrophotometric and extraction-based micro-method for Pd determination. It is reported that assessment of quality control in Pd catalysts can be achieved by a cheap soft drink additive dye (yellow-6). This is mentioned further that in the pH range of 3–9.5 the dye Y6 forms a red complex with Pd(II) which has λ_{max} at 553 nm. The reported pH is 7.0 and themolar absorptivity of 1.14×10^4 L/mol cm is achieved. The complex has a 1:2 composition and stability constant is 3.92×10^{11}. The complex so formed is extracted with cationic surfactant Aliquat 336 using Me isobutyl ketone as solvent. A number of parameters like effect of pH, ionic strength, organic phase volume, Y6 and concentration of complex formation have been studied. The micro-methods have been successfully applied for the determination of Pd(II) in catalyst samples using commercial drinking soda having Y6 dye.

6.70   2-Acetylthiopyran Thiocyanate (1-Thiopyran-2-yl)ethanone Thiocyanate (ATPT)

Borhade [70] has reported synthesis of a new reagent ATPT for determination of Pd(II) contents. It is reported that Pd(II) forms a yellow-coloured complex with ATPT which is extractable into n-hexane at pH 6.0. The absorbance maximum for this complex is at 520 nm. The composition of the complex is 1:2 based on thermal study. The author claims that this method permits separation of palladium from palladium catalysts and other binary mixtures and allows Pd determination successfully.

References

1. Rancic, S.M., Mandic, S.D.N., Bojic, A.L., Veljkovic, S.M.D., Zarubica, A.R. and Jankovic, P.L. Application of the reaction system methylene B-$(NH_4)_2S_2O_8$ for the kinetic spectrophotometric determination of palladium in citric buffer media. *Hem. Ind.*, 2017, 71(2), 97–104.
2. Azmi, S.N.H., Iqbal, B.H.K.A., Sayabi, S.A.M.A., Quraini, N.M.K.A. and Rahman, N. Optimized and validated spectrophotometric method for determination of palladium (II) in synthetic mixture and automobile workshop area samples. *J. Assoc. Arab Univ. Basic Appl. Sci.*, 2016, 19, 29–36.
3. Keyvanfard, M., Rezaei, B. and Alizad, K. Ultra-trace determination of palladium (II) by spectrophotometric flow injection analysis. *Anal. Bioanal. Chem. Res.*, 2017, 4(1), 11–20.
4. Madan, P. and Barhate, V.D. Extractive and spectrophotometric determination of palladium (II) using 2-(5-bromo-2-oxoindolin-3-ylidene)hydrazine carbothioamide as an analytical reagent. *J. Chem. Pharm. Res.*, 2015, 7(12), 227–233.
5. Kuchekar, S., Naval, R. and Han, S.H. Development of a reliable method for the spectrophotometric determination of palladium (II) with o-methoxyphenyl thiourea: separation of palladium from associated metal ions. *S. Afri. J. Chem.*, 2014, 67, 226–232.
6. Dhanavate, S.D., Garole, D.J., Garole, V.J., Tetgurue, S.R. and Sawnt, A.D. Extractive spectrophotometric determination of Pd(II) with sodium sat of hexamethyleneimine carbodithioate. *Int. Lett. Chem. Phys. Astron.*, 2013, 23, 20–28.
7. Mathew, B., Mini, V. and Deepthi, B. Spectrophotometric determination of palladium (II) using a nitrogen sulfur and oxygen donor triazene. *Orient. J. Chem.*, 2010, 26(1), 233–238.
8. Mathew, B. and Innocent, D. Spectrophotometric determination of palladium (II) using thioglycolic acid. *Asian J. Chem.*, 2010, 22(10), 7551–7556.
9. Shemirani, F., Kozani, R.R., Jamali, M.R., Assadi, Y. and Hosseini, M.R.M. Cloud-point extraction, preconcentration and spectrophotometric determination of palladium water samples. *Int. J. Environ. Anal. Chem.*, 2006, 86(14), 1105–1112.
10. Rahman, G.M., Mizanur, H.M., Mustafa, A.I. and Hossam, M.A. Spectrophotometric determination of palladium (II) with rubeanic acid (RA) in presence of ethylene diamine (EDA), diethylenetriamine (DETA), dimethylsulfoxide (DMSO) and ammonium thiocyanate (NH_4SCN). *Dhaka Univ. J. Sci.*, 1998, 46(2), 175–183.
11. Guo, Z., Yuan, M. and Zhang, S. Study on the spectrophotometric determination of palladium using 2,6,7-trihydroxy-9-(3,5-dibromo-4-hydroxy) phenylfluorone and cetyltrimethylammonium bromide. *Fenxi Shiyanshi*, 1995, 14(5), 18–21.

12. Dias, L.F. and Nozaki, J. Spectrophotometric determination of Pd(II) with p-dimethylaminebenzylidinerhodanine. *Braz. Arch. Biol. Technol*, 1999, 42(2).
13. Ge, A. and Pan, J. Spectrophotometric determination of palladium with 5(5-nitro-2-hydroxyphenylazo)rhodanine. *Yejin Fenxi*, 2001, 21(3), 24–25.
14. Ma, D., Li, Y., Lu, Q. and Wang, Y. Spectrophotometric determination of palladium (II) with new reagent N-octyl-N'-(sodium p-aminobenzenesulfonate)thiourea (OPT). *J. Chinese Chem. Soc. (Taipei, Taiwan)*, 2001, 48(68), 1111–1114.
15. Brujjoniti, M.C., Mucchino, C., Tarasco, E. and Sarzanini, C. On-line preconcentration and ion-chromatographic separation and spectrophotometric determination of palladium at trace level. *J. Chromatogr. A.*, 2003, 1007(1–2), 93–100.
16. Kumar, A., Sharma, P., Chandel, L.K. and Kalal, B.L. Synergistic extraction and spectrophotometric determination of palladium (II), iron (III) and tellurium (IV) at trace level by newly synthesized p-[4-(3,5-dimethylisoxazolyl) azophenyl azo]calix (4) arene. *J. Incl. Phenom. Macrocycl. Chem.*, 2008, 61(3–4), 335–342.
17. Vojkovic, V. and Druskovik, V. Simultaneous determination of Pd(II) and Gold (III) in mixtures by third derivative spectrophotometry using 3-hydroxy-2-methyl-1-phenyl-4-pyridone ligand. *Croat. Chem. Acta.*, 2003, 76(1), 87–92.
18. Lokhande, R.S., Nemade, H.G., Chaudhary, A.B. and Hundiwale, D.G. Extractive spectrophotometric determination of palladium (II) using 3-phenoxybenzoldome. *Asian J. Chem.*, 2001, 13(2), 596–602.
19. Eskandari, H. and Khoshandam, M. Sensitive and selective spectrophotometric determination of palladium (II), ion following its preconcentration using modified magnetite nanoparticles and 3-phasic back extraction. *Micro Chim. Acta.*, 2011, 175(3–4), 291–299.
20. Pourreza, N. and Rastegarzadeh, S. Optical test strip for spectrophotometric determination of palladium based on 5-(p-dimethylaminobenzylidene)rhodanine reagent. *Can. J. Anal. Sci. Spectrosc.*, 2004, 49(5), 314–319.
21. Shetty, P., Nityananda, S. and Gadag, R.V. Spectrophotometric determination of palladium (II) using piperonal thiosemicarbazone. *Ind. J. Chem. Technol.*, 2003, 10(3), 287–290.
22. Zhong, Y-H., Huang, Q-L., Zhand, X., Huang, Z.J., Hu, Q-F., Yang, G-Y. Study on solid phase extraction and spectrophotometric determination of palladium with MCI-GEL resin. *Guangpuxue Yu Guangpu Fenxi*, 2007, 27(2), 360–363.
23. Sathe, G.B., Vaidya, V.V. and Ravindra, G. Extractive spectrophotometric determination of palladium (II) using novel Salen ligand. *Int. J. Adv. Res.*, 2015, 3(4), 699–704.
24. Huang, Z., Wang, X.G. and Chang, J. Solidphase extraction and spectrophotometric determination of palladium using 2-methyl-5-(4-carboxyphenylazo)-8-hydroxyquinoline as a chromogenic reagent. *Asian J. Chem.*, 2010, 22(1), 365–372.
25. Salinas, F., Espinosa-Mansilia, A., Lopenz-Martinez, L. and Lopenz-De-Alba, P.L. Selective extraction - spectrophotometric determination of micro-amounts of palladium in catalysts. *Chem. Anal. (Warsaw, Poland)*, 2001, 46(2), 239–248.

26. Shrivas, K., Patel, K.S. and Hoffman, P. Flow injection analysis spectrophotometric determination of palladium. *Anal. Lett.*, 2004, 37(3), 507–516.
27. Gangadharppa, M. and Raveendra Reddy, P. Direct and derivative spectrophotometric determination of palladium with 2-amino acetophenone isonicotinoyl hydrazone (2AAINH). *J. Ind. Chem. Soc.*, 2006, 83(11), 1130–1134.
28. Jabbari, A., Barzegar, M. and Mohammadi, M. Catalytic kinetic spectrophotometric determination of palladium (II) and its application to the determination of traces of propylthiouracil. *Ind. J. Chem. Sect. A: Inorg. Bio-inorg. Phys. Theoret. Anal. Chem.*, 2005, 44A(6), 1215–1218.
29. Shelar, Y.S., Aher, H.R., Kuchekaro, S.R. and Han, S.H. Extractive spectrophotometric determination of palladium (II) with o-methylphenylthiourea from synthetic mixtures. *Bulg. Chem. Commun.*, 2012, 44(2), 172–179.
30. Varghese, A. and Khadar, A.M.A. Highly selective derivative spectrophotometric determination of palladium (II) in hydrogenation catalysts and alloy samples. *Ind. J. Chem. Technol.*, 2011, 18(3), 177–182.
31. Srivani, I., Kumar, A.P., Reddy, P.R., Reddy, K.P.R. and Reddy, V.K. Synthesis of 2-hydroxy-3-methoxy benzaldehyde thiosemi carbazone (HMBATSC) and its application for direct and second derivative spectrophotometric determination of palladium (II). *Annali di Chimica (Rome, Italy)*, 2007, 97(11–12), 1237–1245.
32. Huang, Z., Wang, S., Yang, X., Wei, Q. and Cheng, J. Sold phase extraction and spectrophotometric determination of palladium using hexyl benzimidazolyl sulfide as a chromogenic agent. *Chem. Anal. (Warsaw, Poland)*, 2008, 53(3), 347–356.
33. Yang, B., Zhu, L., Huang, Z., Yang, G. and Yin, J. Guijinshu, Study on the determination of Pd in cyanide slag by SBDR solid phase extraction enrichment separation and spectrophotometry. *Prec. Met.* 2005, 26(1), 39–42.
34. Rao, M.R. and Chandra Sekhar, K.B. Sensitive derivative spectrophotometric determination of palladium (II) using 3,5-Dimethoxy-4-hydroxybenzaldehyde iso nicotinoylhydrazone in presence of micellar medium. *Pharma Chem.*, 2011, 3(2), 358–369.
35. Gaikwad, S.H. and Anuse, M.A. A sensitive extractive spectrophotometric method for determination of palladium (II) with 1,3-bis (hydroxymethyl) benzimidazole-2-thione in catalysts. *Ind. J. Chem. Technol.*, 2003, 10(5), 447–453.
36. Reddy, S.A., Raman, P.V. and Reddy, A.V. Synthesis of novel analytical reagent, 2,6-diacetylpyridine bis-4-phenyl-3-thiosemicarbazone and its analytical applications. Determination of Pd(II) in spiked samples. *J. Chem. Pharm. Res.*, 2015, 7(8), 146–154.
37. Soliman, A.A., Majid, S.R. and Ahaby, F.A. Spectrophotometric determination of palladium using 2-hydrazinopyridine. *Eur. J. Chem.* 2014, 5(11), 150–154.
38. Qu, W., Wang, Z-M., Zhou, C.Y. and Liu, S. Spectrophotometric determination of palladium ore with 4-(5-chloro-2-pyridine)-azo-1,3-diaminobenzene. *Yejin Fenxi*, 2012, 32(2), 55–58.
39. Yuan, L., Ma, M. and Ren, X. Spectrophotometric determination of palladium using bromo-sulfonazo III as chromogenic reagent. *Lihua Jianyan, Huaxue Fence*, 2006, 42(5), 366–367.
40. Li, Y., Yang, X-Z., Li, X-S., Yao, F-Q. and Hu, Q-F. Study on solid phase extraction and spectrophotometric determination of palladium with MCI GEL CHP 2OY as solvent. *Asian J. Chem.*, 2011, 23(11), 4838–4840.

41. Salvi, P., Sharma, P., Chopra, J., Rathore, M.K., Dashora, R. and Goswami, A.K. Analytical application of 3-hydroxy-3-propyl-1-(4-carbamoyl sulfamoyl)phenyl triazene (CSPT) in the spectrophotometric determination of palladium (II). *J. Appl. Chem. (Lumami, India)*. 2018, 7(2), 443–448.

42. Dong, H-R., Zhang, J-C. and Hu, S-C. Ion floatation spectrophotometric determination of palladium (II) content in acetaldehyde catalyst. *Guang pu Xue Yu Guang Pu Fen Xi Guangpu*. 2004, 24(3), 345–347.

43. Lin, H., Zhu, L., Ai, H., Hu, Q.F. and Yang, G. Spectrophotometric determination of palladium with 2-(2-quinolinylazo)-1,3-diaminobenzene Guijinshu, 2004, 25(4), 63–66.

44. Yang, L-J., Huang, Q-B., Li, M., Li, D-L., Zhu, L-Y. and Yin, J-Y. Spectrophotometric determination of palladium with p-sulfobenzylidenerhodanine. *Gungpu Shiyanshi*. 2004, 21(6), 1151–1153.

45. Huang, Z-J., Hu, Q-F., Wang, J-Q. and Yang, G-Y. Spectrophotometric determination of palladium with 5-iodo[Calix (4) arene] aminoquinoline. *Fenxi Shiyanshi*, 2004, 23(6), 45–47.

46. Chen, W., Ma, W-Z.Q., Xu, G. and Chen, L. Spectrophotometric determination of palladium with 2,4-dichlorophenylfluorone (DCIPF). *Yejin Fenxi*, 2006, 26(4), 95–95.

47. Chen, S., Chen, D., Liang, H., Ge, C. and Pan, F. Spectrophotometric determination of palladium (II) with 2-(5-carboxy-1,3,4-triazolylazo)-5-diethylamino aniline. *Yejin, Fenxi*, 2006, 26(2), 50–52.

48. Shavkunova, M.Y., Nezgovorova, N.N., Sinegribova, O.A., Viasov, A.Y. and Smetannikov, A.F. Spectrophotometric determination of palladium concentration in technological solutions. *Khimicheskaya Technol. (MOSOW, Russian Federation)*, 2009, 10(II), 700–704.

49. Lokhande, R.S., Sawant, A.D. and Barhate, V.D. Extractive spectrophotometric determination of palladium (II) with 2-hydroxy-3-nitro-5-methyl acetophenone oxime (HNMA). *Res. J. Chem. Environ.*, 2007, 11(2), 70–71.

50. Wang, Y. and Liu, B. Study on spectrophotometric determination of palladium with new reagent 4-(2-thiazolyazo)-5-diethylamino-aminobenzene. *Yejin Fenxi.*, 2001, 21(5), 18–20.

51. Tang, M., Liang, A., Liu, Q. and Jiang, Z. Nanocatalytic spectrophotometric determination of palladium (II) using $NiCl_2$-NaH_2PO_2 system. *Guangxi Shifan Daxue Xuebao, Ziran Kexueban*, 2011, 29(3), 47–51.

52. Liu, X., Zhu, X., Lin, A. and Li, X. Spectrophotometric determination of palladium (II) in palladium - chlorophosphanazo-pI-Tween-60 ternary system. *Xiyou Jinshu*, 2002, 26(3), 235–237.

53. Wu, J. and Zhao, Z. Solid phase transmission spectrophotometric determination of palladium using PVC functional membrane. *Linhua Jianyan, Huaxue Fence*, 2002, 38(2), 85–86, 88.

54. Wu, X., Lin, H., Zhu, L., Hu, X. and Yang, G. Study on solid phase extraction and spectrophotometric determination of palladium with 4-hydroxy-1-naphthalrodanine. *Huangjin*, 2004, 25(11), 47–50.

55. Zhu, L.,Hu, Q., Yang, X., Yin, J. and Yang, G. Study on solid phase extraction and spectrophotometric determination of palladium in cyanide residue. *Huaxue Tongbao*, 2005, 68(2), 154–157.

56. Huang, Z., Liu, Y. and Xie, Q. Solid phase extraction and spectrophotometric determination of palladium with 2-(p-carboxylphenylazo) benzothiazole. *Huangjin*, 2007, 28(7), 42–44.

57. Yang, W., Hu, Q., Huang, Z., Yin, J., Xie, G. and Chen, J. Solid phase extraction and spectrophotometric determination of palladium with 2-(2-quinolylazo)-5-diethylaminobenzoic acid. *Cryst. Res. Technol.*, 2006, 41(12), 821–828.
58. Hu, Z-Y. and Luo, D-C. Colouring spectrophotometric determination of palladium (II) with 4-(5-chloro-2-pyridyl)-azo-1,3-diaminobenzene. *Fenxi Shiyanshi*, 2012, 31(1), 116–118.
59. Bagal, M.R., Shaikh, U.P.K., Sakhre, M.A., Vaidya, S.R., Lande, M.K. and Arbad, B.P. Spectrophotometric determination of palladium (II) with N-decylpyridine-4-amine from malonate media. *J. Ind. Chem. Soc.*, 2014, 91(5), 845–851.
60. Zhu, X-D., Ge, C-H., Liang, H-D., Pan, F-Y. and Jiang, R-H. Spectrophotometric determination of palladium (II) with 2-(5-carboxy-1,3,4-triazoly-azo)-5-diethylaminobenzoic acid. *Fenxi Kexue Xuebao*, 2008, 24(2), 230–232.
61. Huang, Z-J., Huang, F. and Xie, Q-Y. Solid phase extraction and spectrophotometric determination of palladium with 1-(2′-benzothiazole-3-(4′-carboxylbenzene)triazene. *Fenxi. Shiyanshi*, 2008, 27(2), 58–61.
62. Wang, B., Qin, S., Ruan, S. and Zhang, M. Spectrophotometric determination of palladium based on liquid-solid extraction separation in Tween-80-$(NH_4)_2SO_4$-PAR system. *Xiyou Jinshu*, 2002, 26(4), 317–320.
63. Koosha, E., Ramezani, M. and Niazi, A. Optimization of air-assisted liquid-liquid micro extraction by Box-Behnken design for spectrophotometric determination of palladium in water samples. *J. Anal. Chem.*, 2019, 74(11), 1073–1080.
64. Narhari, K.K., Prabhulkar, S.G. and Patil, R.M. Extractive spectrophotometric determination of palladium metal at trace level using (p-methoxyphenyl) ethane 1,2-dion-1-oxime as analytical reagent. *Int. J. Chem. Sci.*, 2010, 8(2), 835–845.
65. Tymoshuk, O.S., Fedyshyn, O.S., Oleksiv, L.V., Rydhuk, P.V. and Matiychuk, V.S. Spectrophotometric determination of palladium (II) ions using a new reagent: 4-(N′(-4-imino-2-oxothiazolidine-5-ylidene)-hydrazino)-benzoic acid (p-ITYBA). *J. Chem.*, 2020, 8141853.
66. Quezada, A.A., Noguchi, D., Murikami, H., Ieshima, N. and Sakai, T. Simultaneous injection effective mixing flow analysts system for spectrophotometric determination of palladium in dental alloy and catalyst. *J. Flow Inject. Anal.*, 2015, 32(1), 13–17.
67. Ma, D., Cui, F., Fenglian, H., Wang, Y. and Wang, J. A highly sensitive spectrophotometric reagent for the determination of palladium (II). *Yejin Fenxi*, 2007, 27(8), 14–18.
68. Pasha, C., Eliyas, M. and Stancheva, K. Determination of palladium (II) by direct and derivative spectrophotometric method. *Oxidat. Commun.*, 2020, 43(4), 791–808.
69. Leon-Rodriguez, D., Luis, M., Lopez-Martínez, L. and Lopez-de-Alba, P.L. Making analytical chemistry simpler by rediscovering cheap analytical reagents. Micro Spectrophotometric determination of palladium in catalysts using the food additive dye (Yellow-6). *Microchim. Acta*, 2004, 148(1–2), 49–54.
70. Borhade, S.S. Study of spectrophotometric determination of transition metal using 2-acetyl thiopyran thiocyanate (1-thiopyran-2-yl) ethanone thiocyanate. *World J. Pharm. Pharm. Sci.*, 2017, 6(8), 2437–2445.

section B

---

*Spectrophotometric
Determination of Platinum –
Reagents and Methods*

chapter seven

Spectrophotometric Determination Methods for Platinum

Although the exhaust gases from automobile equipped with catalytic converters result in removal of close to 90% of CO, unburned hydrocarbons, nitrogen oxides, etc., the next challenge to environment is in the form of airborne particulate matter. This particulate matter is generated due to deterioration of catalysts being used. This contains mostly two precious metals palladium and platinum and some rhodium. This new source of pollution needs sensitive methods for determination of trace amounts of platinum in any such sample. A very critical review on this problem has been published by Barefoot [1]. The platinum metal and its congeners form highly coloured complexes with a number of inorganic and organic reagents, so there are a number of methods available to determine them, one being spectrophotometry. The following sections describe various spectrophotometric methods and reagents reported in the recent past for determination of platinum. The methods are brief, informative and provide choice for selection on the basis of individual requirement.

The methods being presented are N, N–O, N–S or donor-based spectrophotometric ones, yet the number of reagents in each category does not make enough quantum for individual chapter. This makes sense if the entire contents are presented as such without making any classified chapters. Thus the methods are being described on the basis of reagents used/method developed.

7.1 Carbon–Polyurethane Powder

Development of a chelating sorbent has been reported by Moawed et al. [2]. The material is prepared by linking carbon nanofiber with ion exchange polyurethane foam (CNF-PUFIX). The paper reports sorption properties of Pt(IV) on the CNF-PUFIX at pH 4–5 and using this spectrophotometric determination has been done, whereas detection limit of 2.4 µg/L, RSD=3.8 for n=6 is achieved. It is also reported that the accuracy

DOI: 10.1201/9781003276418-10

of the procedure has been verified by analytical standard reference materials including pharmaceutical and granite samples.

7.2   Leucoxylene Cyanol FF

Revanasiddappa et al. [3] have proposed a new, highly sensitive yet simple and rapid spectrophotometric method for Pt(IV) determination using leucoxylene cyanol FF. It is reported that the oxidation of leucoxylene cyanol FF results in a blue form of xylene cyanol FF by Pt(IV) in H_2SO_4 medium of pH 1.0–2.5. This form of dye shows an absorption maxima at 620 nm in an acetate buffer medium of pH 3.0–4.5. The system obeys Beer's law in the concentration range of 0.3–2.5 μg/mL. The reported molar absorptivity as well as Sandell's sensitivity values are 5.1×10^4 L/mol cm and 0.0038 μg/cm^3, respectively. The application of the method has been successfully done to diverse samples of Pt contents such as pharmaceuticals, soil, natural water, plant material, catalysts and synthetic alloys.

7.3   1-3-Dimethyl-2-Thiourea and Bromocresol Green

Terada et al. [4] have proposed an extraction-based spectrophotometric method for determination of Pt(II). The authors have studied the extraction behaviour of Pt(II) with 1,3-dimethyl-2-thiourea from a chloride solution. The method incorporates bromocresol green as a counter anion and 1,2-dichloroethane as an extraction solvent. The system for Pt(II) absorbs at 413 nm, (λ_{max}), so the extraction efficiency has been measured at this wavelength. The quantitative extraction time for Pt(II) with 1,2-dichloroethane is within 15 minutes and parameters like pH effect, DMTU concentration, chloride concentration on the extraction have also been reported. The pH range reported is between 3.3 and 4.2. It is mentioned that sulphate, nitrate, perchlorate, cobalt (II) and nickel (II) do not interfere in the determination. However, presence and interference of Mn(II), Cu(II), Zn(II), Pd(II), Ag(I) and Cd(II) can be masked by extraction separation with 1,5,9,13-tetra thiacyclohexadecane (TTCH) in the presence of EDTA as reported.

7.4   Astrafloxin FF

Bazel et al. [5] have studied and reported optimum conditions for extraction of the ion associates of platinum using thiocyanate ions and polymethine basic dye astrafloxin FF by aromatic hydrocarbons and acetic acid esters. A considerable increase in the extraction of the platinum ion associates is achieved by introducing water-soluble donor active solvents, viz. hexamethylphosphoric triamide, *N,N*-diethylacetamide, DMF and *N,N*-diethylformamide. This simultaneously suppresses extraction of simple dye salt. The spectrophotometric parameter, e.g., molar absorptivity of extract, is high – 8.1–13.3×10⁴, and it is claimed that 84%–96% platinum is extracted by single extraction. Thus, the procedure converts platinum into ion associates, finally separating platinum from many other elements, viz. Cu, Cd, Ni, Co, Cr, Pb, In, As, Pd, Ir, Rb and Ru. It is reported that Pt(II) is extracted as an ion associate under standard conditions whereas Pt(IV) can be extracted only after thermal treatment. It is claimed to be the new procedure which determines trace quantities of Pt(II) and Pt(IV) using extraction spectrophotometry.

7.5 Piperonal Thiosemicarbazone

Shetty et al. [6] have proposed a rapid, selective, sensitive yet simple method for spectrophotometric determination of Pt based on colour reaction of Pt(IV) with piperonal thiosemicarbazone (PATS). It is reported that in 0.008–0.32 M H_2SO_4 medium a greenish yellow complex is formed which has an absorption maximum at 360 nm. Beer's law is obeyed upto 6.5 ppm of Pt and the optimum concentration range is 1–5.1 ppm of Pt. The reported values of molar absorptivity and Sandell's sensitivity are 3.239×10⁴ L/mol cm and 0.006 µg/cm², respectively. It is mentioned that this method was used for the determination of Pt in hydrogenation catalysts and Pt complexes.

7.6 4-[N,N-(Diethyl)amino]benzaldehyde Thiosemicarbazone (DEABT)

Naik et al. [7] have proposed an analytical reagent 4-[*N,N*-(diethyl)amino] benzaldehyde thiosemicarbazone for the spectrophotometric determination of Pt(IV). It is reported that the mentioned reagent DEABT forms a 1:2 yellow complex, which is sparingly soluble in water but is completely soluble in the medium of water-ethanol-DMF mixture. This complex has λ_{max}=405 nm, whereas Beer's law is obeyed up to 7.80 µg/cm³. The reported concentration range of determination is 0.48–7.02 µg/cm³. Further, molar absorptivity and Sandell's sensitivity values as reported are 1.755×10⁴ L/mol cm and 0.0012 µg/cm², respectively. The value of relative error and coefficient of variation (n=6) for the method are ±0.43% and 0.35%. Features of the method highlighted include rapidity and solubility in water ethanol–DMF medium, no time-consuming extraction process and no interference of associated metal ions. Versatility of method is evidenced by the application of this to determination of Pt in environmental, pharmaceutical, alloy, catalyst and complex samples.

7.7   N-(3,5-Dimethylphenyl)-N'-(4-Aminobenzenesulfonate) thiourea (DMMPT)

Using rapid reaction of platinum (IV) with N-(3,5-dimethylphenyl)-N'-(4-aminobenzenesulfonate) thiourea and its subsequent solid-phase extraction with C_{18} membrane, Xin et al. [8] have proposed a method for spectrophotometric determination of Pt(IV). It is reported that in pH 3.8 buffer and cetyl trimethyl ammonium bromide (CTMAB) medium, the reagent DMMPT forms a violet complex in the ratio of 1:3 (Pt:DMMPT). The complex is enriched by solid-phase extraction using C_{18} membrane with an enrichment factor of 200. The absorbance at 755 nm, in the Beer's law range facilitates determination of Pt(IV). The molar absorptivity reported is 9.51×10^4 L/mol cm. RSD value for 11 replicates is 1.79% at 0.01 µg/mL level. The detection limit of 0.02 µg/L is achieved using this method and values obtained agree well with those of ICP-MS method, making the method equally competent.

7.8   1-Phenyl-4-Ethylthiosemicarbazide (PETS)

Al-Attas [9] has developed a simple, highly sensitive as well as selective spectrophotometric method for determining Pt(IV) using 1-phenyl-4-ethylthiosemicarbazide as a complexing ligand. It is reported that at pH 3.0, the named reagent forms a greenish complex with Pt(IV) in a mole ratio of Pt(IV):PETS (1:2), which floats quantitatively with oleic acid surfactant. The maximum absorbance of the complex is 715 nm in both aqueous as well as surfactant layer, thus facilitating its determination. Beer's law obeyed is up to 0.39 mg/L concentration for aqueous solution of Pt(IV). The molar absorptivity reported is 0.14×10^5 and 0.5×10^5 L/mol cm for aqueous and surfactant, respectively. It is mentioned as a successful method for Pt(IV) analysis is natural waters, prepared solid complexes and simulated samples. The results so obtained are comparable to those obtained using AAS.

7.9   Phenanthrenequinonemonosemicarbazone (PQSC)

Rai et al. [10] have developed a method on the basis of reaction of PQSC with Pt(IV). It is mentioned that the given reagent forms a brown precipitate with Pt(IV), which can be dissolved in DMF, although it is partially soluble in ethanol and chloroform. Since the solution has an intense and stable colour, the reaction is used for spectrophotometric determination of Pt metal as mentioned.

7.10   Anthranilic Acid

Nambiar et al. [11] have developed a rapid and sensitive method for spectrophotometric determination of Pt(IV) using anthranilic acid as a ligand. The ligand reacts with Pt(IV) at pH 3.0 to give a pink complex showing λ_{max} at 372 nm. The reported molar absorptivity and Sandell's sensitivity values are 3.41×10^4 L/mol cm and 5.6×10^{-3} µg/cm², respectively. The method reported includes interference studies and application of synthetic samples as mentioned.

7.11   Flow Injection Analysis Spectrophotometric Determination of Platinum

Patel et al. [12] have devised a new simple, rapid yet selective method for the flow injection analysis (FIA) spectrophotometric determination of platinum. The method reported is based on colour reaction of Pt(IV) with $SnCl_2$ in HCl medium. Cetylpyridinium chloride and Triton X-100 are used as mixed surfactants in order to enhance the sensitivity of the method. The λ_{max} as reported is 405 nm. The molar absorptivity value is 3.0×10^3 L/mol cm. The reported detection limit of the method is 150 ng/mL. The optimum concentration range for the determination of Pt is 0.5–18 µg/mL. The other factors reported are slope 0.0086, intercept −0.001 and correlation coefficient +0.99. It is reported that the method was tested for Pt analysis in catalytic samples.

7.12   Isonitroso p-Methyl Acetophenone Phenylhydrazone (HIMAPH)

A new reagent isonitroso p-methyl acetophenone phenyl hydrazone has been synthesized and used for extractive spectrophotometric determination of Pt(IV) by Kumar et al. [13]. It is reported that the named reagent quantitatively extracts (99.17%) into toluene from an aqueous solution of pH 5.0–7.2. The method requires presence of 1 mL of 2 M sodium acetate followed by digestion on boiling water bath for about 35–40 minutes. The ligand HIMAPH forms a reddish yellow-coloured complex which is extracted into toluene. The extracted complex shows λ_{max} at 465 nm in the Beer's law concentration range of 0.1–1.2 µg/mL. The authors also report composition of the complex as 1:4 (Pt:HIMAPH) by two different methods and molar absorptivity value as 10,731.71 L/mol cm. The method is useful even in the presence of various ions and Pt(IV) determination in silver alloy or Pt catalyst samples.

7.13 Potassium Bromate and 2-(4-Chloro-2-phosphonophenyl) azo-7-(2,4,6-trichlorophenyl) Azo-1,8-dihydroxy-3,6-naphthalene Disulfonic Acid

Kou et al. [14] have reported use of the mentioned reagents potassium bromate and 2-(4-chloro-2-phosphonophenyl)azo-7-(2,4,6-trichlorophenyl) azo-1,8-dihydroxy-3,6-naphthalene disulfonic acid using spectrophotometry for Pt(IV) determination. The range of concentration in which the method works is 1.0–25 µg/M. The catalytic spectrophotometric method has a relative standard deviation of 2.8%–3.6% with a 99.8%–100.5% recovery. Application for determination of trace Pt in Pt-Rh catalysts and Pt-Pd ore is reported.

7.14 1-(2-Pyridylazo)-2-naphthol

Dong et al. [15] have developed an effective method for spectrophotometric determination of Pt(IV) using 1-(2-pyridylazo)-2-naphthol and molten naphthalene as diluent. The method reports formation of a green complex when Pt(IV) acts on PAN at 90°C. The complex so formed can be quantitatively extracted into molten naphthalene. Further, this is anhydrously dissolved in $CHCl_3$ for determination at 690 nm against the reagent blank in the Beer's law concentration range of 0.5–5.0 µg/mL. The molar absorptivity and Sandell's sensitivity values are 1.6×10^5 L/mol cm and 0.0048 mg/cm^2, respectively. Even in the presence of various ions, the method is successful for Pt(IV) determination and can be applied to synthetic samples as well.

7.15 2-(5-Bromo-2-oxoindolin-3-ylidene) hydrazine Carbothioamide (HBITSC)

A simple, rapid yet sensitive spectrophotometric method for Pt(IV) determination has been proposed by Barhate et al. [16] using HBITSC as an analytical reagent. It is reported that the synthesized reagent quantitatively extracts Pt(IV) up to 99.67% into n-amyl alcohol from an aqueous solution in the pH range of 3.0–6.0 in 3 mL of acetic acid sodium acetate buffer (4.6 pH). An intense peak at 505 (λ_{max}) is shown by this n-amyl alcoholic extract. The Beer's law range of detection is 1.0–9.0 µg/cm^3 for

Pt(IV). The reported molar absorptivity and Sandell's sensitivity values are 8452.53 L/mol cm and 23.07 ng/cm^2 for a 1:2 complex. The proposed method has been tested for interference of various ions and is rapid, sensitive and reproducible as reported by the authors.

7.16   *o-Methylphenylthiourea and Iodide*

Kuchekar et al. [17] have reported a new spectrophotometric method for the determination of Pt(IV) using o-methylphenylthiourea (OMPT). It is reported that platinum (IV) is determined as its ternary complex in 1:1:2 (Pt (IV):OMPT:iodide) as a yellow-coloured complex after liquid–liquid extraction from aqueous KI media (0.1 mol/L) using 4.5×10^{-3} mol/L OMPT in 10 mL chloroform. The complex is measured at 362 nm, in the Beer's law range up to 15 µg/mL. The reported molar absorptivity and Sandell's sensitivity values are 1.25×10^4 L/mol cm and 0.016 µg/cm^2, respectively. RSD value as reported is 0.34 (n = 10). Interference of various ions has also been studied and the method is applicable to Pt(IV) determination in binary synthetic as multicomponent, Adam catalyst and pharmaceutical samples. The method has been validated by comparing the results obtained by AAS. Further, it is proposed that liquid–liquid extraction may be useful for separation of Pt(IV), Pd(II) and Ni(II) together.

7.17   *o-Aminobenzylidenerhodanine*

Huang et al. [18] have developed a method for spectrophotometric determination of Pt using o-aminobenzylidenerhodanine (ABR) as reagent. The method uses digestion of sample with microwave and subsequent determination. In HCl medium, ABR reacts with platinum to form a 2:1 complex, whose molar absorptivity is 7.7×10^4 L/mol cm at 525 nm. The applicable Beer's law range is 0–50 mg/25 mL. The method has been applied to determine Pt in cyanide residues and catalyst samples successfully.

7.18   *Microwave Digestion and Spectrophotometric Determination of Platinum*

Yang et al. [19] have reported a method for Pt(IV) determination using microwave digestion and p-aminobenzylidinerhodanine (ABR) as a reagent. The reagent p-aminobenzylidenerhodanine reacts with Pt(IV) in HCl solution giving a 2:1 stable complex. At 525 nm the complex absorbs and obeys Beer's law range of 0–50 µg/25 mL. The reported molar absorptivity is 5.98×10^4 L/mol cm. The method can be successfully applied to the determination of Pt in catalyst samples.

7.19   2-Hydroxy-1-naphthalrhodanine

Lin et al. [20] report synthesis and application of 2-hydroxy-1-naphthal-rhodanine in the spectrophotometric determination of Pt. It is reported that the mentioned reagent in the presence of HCl solution and Tween-80 medium reacts with Pt to give a stable 2:1 complex. This complex absorbs at 540 nm and has molar absorptivity of 1.02×10^5 L/mol cm. The system obeys Beer's law in the concentration range between 0 and 2 μg/mL. It is reported as a successful method for Pt determination in catalysts.

7.20   p-Sulfobenzylidenerhodanine (SBDR)

The named reagent SBDR has been used by Li-Ming et al. [21] for Pt(IV) determination spectrophotometrically. The reagent gives in the presence of HCl solution a 2:1 stable complex with platinum which has a λ_{max} 545 nm. The reported Beer's law range is 0–50 μg/25 mL and molar absorptivity is 7.25×10^4 L/mol cm. The method has been successfully applied to the determination of Pt in catalysts.

7.21   5-(H)-Acidazorhodanine

Qiao et al. [22] have explored colour reaction of platinum with 5-(H)-acidazorhodanine (HAR) for spectrophotometric determination. It is reported that in the presence of HCl the reagent HAR reacts with platinum to give a stable 2:1 complex. The wavelength reported for absorbance is 540 nm in the Beer's law range up to 0–30 μg/25 mL. The molar absorptivity is 1.06×10^5 L/mol cm as reported by the authors. The method can be applied successfully for the determination of platinum in catalysts.

7.22   5-(2-Hydroxy-4-sulfo-5-chlorophenol-
1-azo)thiorhodanine (HSCT)

Bali et al. [23] have developed a method for spectrophotometric determination of Pt using HSCT. The reagent gives colour reaction in the presence of HCl with Pt to give a 2:1 stable complex. It is measured at 535 nm in the Beer's law range of 0–50 μg/25 mL and the complex can be determined spectrophotometrically. The molar absorptivity is 6.24×10^4 L/mol cm and method is useful for Pt determination in catalysts.

7.23 2-Hydroxy-5-sulfobenzenediazoamino azobenzene (HSDAA)

A method has been proposed by Li et al. [24] using chromogenic reaction between Pt(IV) and 2-hydroxy-5-sulfobenzenediazoaminobenzene in the presence of CPB/OP micro-emulsion. It is reported that Pt(IV) reacts with HSDAA in the presence of CPB/OP micro-emulsion to give a stable yellow – 1:2 complex in NaOH solution of pH 11.7. The complex absorbs at 434 nm in the Beer's law range of 0.04–10 μg/10 mL of Pt(IV). Reported molar absorptivity is 1.59×10^5 L/mol cm and detection limit is 1.14×10^{-6} g/L. RSD value is 1.9%–2.2% and recovery of 98.4%–103.5% is also mentioned with comparable results using AAS, and the method can be used for Pt determination of Pt(IV) in catalysts.

7.24 Tetrabromofluorescein and Rhodamine 6G

Liu et al. [25] have developed a floatation spectrophotometric method of determining Pt with isochromatic dye ion-pair between tetrabromo-fluorescein and rhodamine 6G. [(2R6G)(PtBr$_6$).3(R6G.Br) is resolved and rhodamine 6G (R6G) was made to enter into aqueous phase, when HAc–NaAc buffer of 5.5 was added to [(2R6G)(PtBr$_6$).3(R6G.Br)] in toluene. It is reported that R6GTBF floated by tetrabromofluorescein (TBF) with toluene. After dissolving the float in acetone, the measurements are done at 530 nm. The reported molar absorptivity is 9.94×103 L/mol cm, in the Beer's law range of 0–0.5 mg/L. The method is useful for Pt determination in Pt catalysts.

7.25 2-(5-Iodine-2-pyridylazo)-5-dimethylaminoaniline (5-I-PADMA)

Huo [26] reported a method for spectrophotometric determination of platinum (IV) in catalysts using 2-(5-iodine-2-pyridylazo)-5-dimethyl-aminoaniline as a reagent. In the pH range of 3.7–5:6, 5-I-PADMA forms a stable complex with Pt showing maximum absorbance at 625 nm. The reported molar absorptivity is 4.55×10^4 L/mol cm in the Beer's law range of 0–1.4 μg/mL.

7.26 Polyamide Resin

Liu et al. [27] have developed a preconcentration- and separation-based spectrophotometric determination method using polyamide resin. It is reported that polyamide resin has a high adsorption capacity on Pt and Pd in 0.001–35 mol/L of HCl solution with a saturation adsorption capacity of 20.92 (Pt) and 19.20 (Pd) mg/g. Subsequently adsorbed platinum can be eluted by 50 g/L of acid thiourea solution with pH 1–3 and 60°C–90°C, which is determined using DDO spectrophotometry. The reported method has been applied to preconcentration, separation and determination of trace Pt in mineral samples, whose results agree with those obtained using ICP-AES. The relative error reported is 5.56% for Pt.

7.27   3,4-Diaminobenzoic Acid

Dong et al. [28] have proposed an effective method for Pt determination using colour reaction of Pt with 3,4-diaminobenzoic acid with molten naphthalene as diluent. This is reported that a green complex of Pt-DBA is formed at 90°C, which is extracted into molten naphthalene at pH between 10.0 and 12.5. This organic phase is dissolved in $CHCl_3$ and determined spectrophotometrically. Reported absorbance of the extracted complex is 715 nm against reagent blank. The obeyance of Beer's law is in the range between 0.1 and 2.0×10^{-6}, with a molar absorptivity and Sandell's sensitivity of 1.2×10^6 L/mol cm and 0.00041 mg/cm^2, respectively. The reported method applies well to synthetic samples for Pt determination.

7.28   2-(p-Carboxyphenylazo) benzothiazole (CPABT)

A method for spectrophotometric determination of platinum using solid-phase extraction with 2-(p-carboxyphenylazo)benzothiazole has been proposed by Huang et al. [29]. The method is based on complexation of Pt(IV) with CPABT and subsequent solid-phase extraction with C_8 cartridge. This is facilitated in the presence of Tween-80 and HAc–NaAc buffer solution (pH 3.8–5.5), when a stable 1:1 complex is formed. The coloured complex is extracted by C_8 cartridge and eluted with DMF. The absorbance of this complex is at 508 nm in the concentration range of 0.1–1.2 µg/mL for Pt. Molar absorptivity reported for this is 2.29×10^5 L/mol cm.

7.29   4,5-Dibromo-2-nitrophenylfluorone

Li et al. [30] have established a spectrophotometric determination method for Pt(IV) using chromogenic reaction with 4,5-dibromo-2-nitrophenylfluorone (DBON-PF). It is reported that at pH 3.6 (HAc–NaAc buffer) in the presence of CPB/OP micro-emulsion Pt reacts with DBON-PF to give a stable pink complex in 1:2 ratio. The λ_{max} of 476 nm is suitable to determine this complex in the Beer's law range of 0–14 µg/10 mL of Pt(IV). The molar absorptivity is 1.88×10^5 L/mol cm with detection limit of 1.06×10^{-6} g/L. The application to catalyst samples has recovery of 97.6%–105% and RSD of 1.1%–2.1% is also achieved.

7.30 N-(3,5-Dihydroxyphenyl)-N'-(4-Aminobenzenesulfonate)thiourea

Li et al. [31] have proposed a new highly sensitive, selective yet rapid method for the determination of Pt(IV) based on rapid reaction of DHPABT-Pt(IV). The complex so formed is subsequently extracted with solid-phase method, using C_{18} cartridge. It is described that DHPAB-Pt(IV) violet complex of 1:3 (Pt:DHPABT) is enriched by solid-phase extraction (in 0.05–0.5 mol/L HCl medium) with polymer-based C_{18} cartridge with 200 as enrichment factor. The extracted complex absorbs at 760 nm, in the Beer's law range of 0.01–3.0 µg/L and has molar absorptivity of 1.01×10^5 L/mol cm. RSD for 0.01 µg/mL solution is 1.86% (11 replicates). Further, the detection limit reaches 0.02 µg/L in the original samples. The method was applied to samples of Pt(IV) in H_2O and soil and results are also comparable to results obtained using ICP-MS method as reported.

7.31 5-(5-Iodo-2-pyridylazo)-2,4-diaminotoluene (5-I-PADAT)

Yang et al. [32] have studied colour reaction of 5-(5-iodo-2-pyridylazo)-2,4-diaminotoluene (5-I-PADAT) with platinum (IV) to establish spectrophotometric method for its determination. It is mentioned that in HAc–NaAc buffer solution of pH 5.5, platinum (IV) forms a stable complex with the ligand. Further, this in 3.5 mol/L HCl solution changes its colour which has higher absorption characteristic and λ_{max} of 595 nm. The reported molar absorptivity of this complex is 3.2×10^4 L/mol cm in the Beer's law range of 0–1.5 µg/L of Pt. The method has been said to be convenient, fast and also has good selectivity.

7.32 p-Rhodanineazobenzoic Acid

Huang et al. [33] have reported a spectrophotometric method combined with solid-phase extraction using p-rhodanineazobenzoic acid (RABA) as reagent. The Pt in catalysts can be determined when in the presence of Na dodecylbenzenesulfonate (SDBS) and KOH-C_6H_{42} HK buffer solution of pH 4.0–6.2, the reagent RABA reacts with Pt(IV). A 1:1 stable complex is formed which is enriched by AccuBond C_8 cartridge and eluted with ethanol. At 520 nm, this ethanolic complex has a molar absorptivity value of 2.49×10^5 L/mol cm. The Beer's law range is 0.1–1.0 µg/mL of Pt. As reported, the results agree well with results obtained by AAS, and RSD of 1.9%–2.3% (n=7) with recovery of 97%–101% is achieved.

7.33   4-(2'-Furalideneimino)-3-methyl-5-mercapto-1,2,4-triazole in n-Butanol

Gaikwad [34] has reported a selective extraction-based spectrophotometric method for determination of Pt(IV) in real samples. The reagent 4-(2'-furalideneimino)-3-methyl-5-mercapto-1,2,4-triazole (FIMMT) reacts with Pt(IV) to form a red complex which forms an aqueous solution of pH 5.4 and is extracted into n-butanol. This complex so extracted can be measured at 510 nm against a reagent blank. The reported molar absorptivity and Sandell's sensitivity are 1.1686×10^4 L/mol cm and 0.017 µg/cm², respectively. The concentration obtained from Ringbom's plot is 3.16–14.79 ppm. The method is rapid, highly selective and can work in the presence of diverse ions. The synthetic mixtures of alloys or even anticancer drugs can be analysed for Pt(IV) as claimed.

References

1. Barefoot, R.R. Determination of platinum at trace levels in environmental and biological materials. *Environ. Sci. Technol.*, 1997, 31(2), 309–314.
2. Moawed, E.A., Ishaq, I., Abdul Rahman, A. and El-Shahat, M.F. Synthesis, characterization of carbon-polyurethane powder and its application for separation and spectrophotometric determination of platinum in pharmaceutical and ore samples. *Talanta*, 2014, 121, 113–121.
3. Revanasiddappa, H.D. and Kumar, T.N. A highly sensitive spectrophotometric determination of platinum (IV) using leuco xylene cyanol FF. *Anal. Bioanal. Chem.*, 2003, 375(2), 319–323.
4. Terada, Y., Harada, A., Saito, K., Murakami, S. and Muromatsu, A. Solvent extraction of platinum (II) with 1,3-dimethyl-2-thiourea and Bromocresol green. *Bunseki Kagaku*, 2003, 52(9), 725–729.
5. Bazel, Y.R., Kulakova, T.A., Studenyak, Y.I., Serbin, R., Rednik, S. and Andruch, V. Extraction of platinum with Astrafloxin FF from aqueous organic solutions: separative extraction spectrophotometric determination of platinum (II) and platinum (IV) species. *J. Anal. Chem.*, 2012, 67(6), 519–526.
6. Shetty, P., Shetty, A.N. and Gadag, R.V. Rapid spectrophotometric determination of platinum (IV) using piperonal thiosemicarbazone. *Ind. J. Chem. Sect. A: Inorg. Bioinorg. Phys. Theoret. Anal. Chem.*, 2002, 41A(5), 988–990.
7. Naik, P.P., Karthikeyan, J. and Shetty, A.N. Spectrophotometric determination of platinum (IV) in alloys, environmental and pharmaceutical samples using 4-[N, N-(diethyl)amino] benzaldehyde thiosemicarbazone. *Environ. Monit. Assess.*, 2010, 17(1–4), 639–649.

8. Xin, Z., Zhou, Y., Huang, Z., Hu, Q., Chen, J. and Yang, G. Study of solid phase extraction prior to spectrophotometric determination of platinum with N-(3,5-dimethylphenyl)-N'-(4-aminobenzenesulfonate)-thiourea. *Microchim. Acta.*, 2006, 153(3–4), 187–191.
9. Al-Attas, A. Separation and spectrophotometric determination of platinum (IV) in natural water, simulated samples and prepared solid complexes using 1-phenyl-4-ethylthiosemicarbazide. *Jordan J. Chem.*, 2007, 2(21), 183–197.
10. Rai, H.C. and Chakraborti, I. Spectrophotometric determination of platinum (IV) with phenahthrenequinonemonosemicarbazone. *Asian J. Chem.*, 2003, 15(3&4), 1687–1692.
11. Nambiar, C.H., Raghvan, Narayan, B., Sreekumar, N.V., Nazareth, R.A. and Bhat, N.G. Spectrophotometric determination of platinum (IV) using anthranilic acid. *J. Ind. Chem. Soc.*, 2002, 79(9), 778–779.
12. Patel, K., Jaiswal, N., Sharma, P. and Hoffman, P. Flow injection analysis spectrophotometric determination of platinum. *Anal. Lett.*, 2006, 39(1), 197–205.
13. Kumar, A., Gupta, S. and Barhate, V.D. Extractive and spectrophotometric determination of platinum (IV) using isonitroso p-methyl acetophenone phenyl hydrazone. *J. Ind. Council Chem.*, 2009, 26(1), 70–73.
14. Kou, M., Kou, F. and Kou, Z. Catalytic spectrophotometric determination of platinum (IV) using potassium bromate and 2-(4-chloro-2-phosphonophenyl)azo-7-(2,4,6-trichlorophenyl)azo-1,8-dihydroxy-3,6-naphthalene disulfonic acid. *Fenxi Huaxue.* 2004, 32(8), 1120.
15. Dong, Y., Gai, K. and Gong, X.X. Spectrophotometric determination of platinum after solid, liquid extraction with 1-(2-pyridylazo)-2-naphthol at 90°C. *Rare Metals (Beijing, China)*, 2004, 23(3), 197–202.
16. Barhate, V.D. and Madan, P.U. Extractive spectrophotometric determination of platinum (IV) using 2-(5-bromo-2-oxoindolin-3-ylidene)hydrazine carbothioamide as an analytical reagent. *World J. Pharm. Pharmaceut. Sci.*, 2016, 5(4), 1939–1947.
17. Kuchekar, S.R., Shelar, Y.S. and Han, S.H. Spectrophotometric determination of platinum (IV) through, the o-methylphenylthiourea and iodide ternary complex after liquid- liquid extraction. *Braz. J. Anal. Chem.*, 2012, 3(10), 421–428.
18. Huang, Q., Zhu, L., Bai, H.W., Xianhua and Yin, J. Spectrophotometric determination of platinum with o-amino benzylidene rhodanine. *Huaugjin*, 2006, 27(4), 52–53.
19. Yang, L-J., Huang, Q-L., Al, H-L.L., H-T. and Zhu, L-Y. Microwave digestion and spectrophotometric determination of platinum. *Guang pu Shiyanshi*, 2004, 21(3), 516–517.
20. Lin, H., Zhu, L., Li, M., Hu, Q. and Yang, G. Study on spectrophotometric determination of platinum with 2-hydroxy-1-naphthalrhodanine. *Guijinshu*, 2005, 26(2), 39–41.
21. Li, M., Li, D., Yang, L., Zhang, Q., Li, H. and Yin, J. Study on spectrophotometric determination of platinum with p-sulfobenzylidenerhodanine. *Huaugjin*, 2004, 25(2), 41–42.
22. Qiao, Y., Dong, X., Yang, G., Zhu, L. and Han, Y. Study on spectrophotometric determination of platinum with 5-(H-acidazo)-rhodanine. *Guijinshu*, 2006, 27(2), 62–64.

23. Bali, H.M., Yang, J-H. and Yin, J-Y. Spectrophotometric determination of platinum with 5-(2-hydroxy-4-sulfo-5-chlorophenol-1-azo) thiorhodanine. *Gungpu Shiyanshi*, 2003, 20(1), 67–69.
24. Li, Y., Ma, W., Xu, X. and Chen, J. Spectrophotometric determination of platinum (IV) with 2-hydroxy-5-sulfobenzenediazoaminobenzeue. *Yejin Fenxi*, 2007, 27(01), 62–64.
25. Liu, B., Liu, Z., Cao, Z. and Gao, J. Floatation spectrophotometric determination of platinum with isochromatic ion-pair between tetrabromofluorescein and rhodamine 6G. *Anal. Chem.*, 2005, 1(1–2), 15–18.
26. Huo, Y-Y. Spectrophotometric determination of platinum (IV) in catalyst with 2-(5-iodine-2-pyridylazo)-5-dimethyl aminoaniline. *Guangpu Shiyanshi*, 2013, 30(3), 1501–1504.
27. Liu, H-L. and Li, S. Spectrophotometric determination of platinum and palladium by preconcentration and separation. *Yang Kuang Ceshi*, 2004, 23(1), 37–39, 43.
28. Dong, Y., Gai, K. and Gong, X. Spectrophotometric determination of platinum after solid-liquid extraction with 3,4-diaminobenzoic acid at 90°C. *Baoji Wenli Xueyuan Xuebao, Ziran, Kexueban*, 2003, 23(3), 189–194.
29. Huang, Z-J., Huang, F., Liu, Y-Y. and Xie, Q-Y. Study on solid phase extraction and spectrophotometric determination of platinum with 2-(p-carboxylphenylazo)benzothiazole. *Fenxi Shiyanshi*. 2008, 27(1), 42–45.
30. Li, Y., Xu, X., Ma, W. and Gao, F. Spectrophotometric determination of platinum (IV) in catalyst with 4,5-dibromo-2-nitro-phenylfluorone. *Yejin Fenxi*, 2008, 28(10), 71–73.
31. Li, Z., Li, X., Zhu, L., Hu, Q., Chen, J. and Yang, G. Solid phase extraction and spectrophotometric determination of platinum (IV) with N-(3,5-dihydroxyphenyl)-N'-(4-aminobenzene sulfonate)thiourea. *Ind. J. Chem. Sect. A: Inorg. Bio-inorg. Phys. Theoret. Anal. Chem.*, 2006, 45A(8), 1852–1855.
32. Yang, X., Yuan, L., Shang, X., Ren, H. and Yang, L. Spectrophotometric determination of platinum (IV) in catalysts with chromogenic reagent 5-(5,iodine-2-pyridylazo)-2,4-diaminotoluene. *Gongye Cuihua*, 2012, 20(6), 70–73.
33. Huang, Z., Liu, Y. and Xie, Q. Solid phase extraction and spectrophotometric determination of platinum in catalysts with p-rhodanineazo benzoic acid. *Yejin Fenxi*, 2007, 27(6), 32–35.
34. Gaikwad, S.H., Lokhande, T.N. and Anuse, M.A. Extraction spectrophotometric determination of micro amounts of palladium (II) in catalysts. *Indian Journal of Chemistry Section a*, 2005, 44(8), 1625–1630.

Index

Note: **Bold** page numbers refer to tables and *italic* page numbers refer to figures.

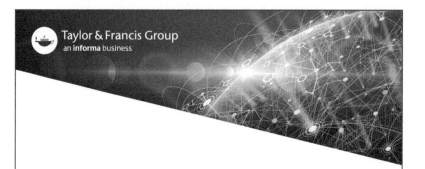